BERRIES OF LABRADOR

Other Books by ELLEN BRYAN OBED

Borrowed Black — A Labrador Fantasy, 1979, 1988 (international edition, 7 languages), 2004 (25th anniversary edtion), 2005 (Inuttitut edition)

Little Snowshoe, 1984, 2014 30th anniversary edition (1st ed. also in Danish)

Wind in my Pocket (poetry), 1990

Wind Dance (poetry), 1999

A Letter from the Snow, 1999

Who Would Like a Christmas Tree?: The Story of the Christmas Tree Farm Around the Year, 2009, 2019

Partridgeberry, Redberry, Lingonberry, Too, 2008

Partridgeberry, Redberry, Tytebaer, Too, 2011

Twelve Kinds of Ice, 2012 (also in Japanese)

BERRIES OF LABRADOR

by ELLEN BRYAN OBED

with watercolours by
VALERIE POWELL

MEMORIAL
UNIVERSITY
PRESS

2024 © Ellen Bryan Obed (text); 2024 © Valerie Powell (artwork); 2024 © Mavis Penney (artwork)

Publisher expressly prohibits the use of *Berries of Labrador* in connection with the development of any software program, including, without limitation, training a machine learning or generative artificial intelligence (AI) system.

Library and Archives Canada Cataloguing in Publication

Title: Berries of Labrador / by Ellen Bryan Obed ; with watercolours by Valerie Powell.
Names: Obed, Ellen Bryan, 1944- author. | Powell, Valerie, illustrator.
Description: Includes bibliographical references and index.
Identifiers: Canadiana 20240447859 | ISBN 9781990445279 (softcover)
Subjects: LCSH: Berries—Newfoundland and Labrador—Labrador.
Classification: LCC SB381 .O23 2024 | DDC 634/.7097182—dc23

Cover art: Valerie Powell
Copy editing: Rebecca Roberts
Cover design, page design, and layout: Alison Carr

Published by Memorial University Press
Memorial University of Newfoundland and Labrador
P.O. Box 4200
St. John's, NL A1C 5S7
www.memorialuniversitypress.ca

Printed in Canada

30 29 28 27 26 25 24   1 2 3 4 5 6 7 8

to the children of Labrador and their grandmothers
who first taught me about their berries
— EBO

We're here in Nain, Labrador, just on the side of Moriah. People of Nain come here in the fall to pick blackberries on this side of the hill. There are other spots where we pick berries like blue and redberries but this is where I normally pick my red and black. It don't really matter to me whether they're big or small. I just enjoy picking berries and I save them in my freezer for the winter to do my baking. Mostly, I just love berry-picking.

I call it my healing garden. I always think of my mother when I berry-pick 'cause that's who taught me how to berry-pick. I used to go with her, but when I used to go berry-picking with her at the beginning when I was younger, I'd just eat — eat all my berries! And not pick in a bucket or a jug. But now, when I go berry-picking, I try not to have a mouthful 'cause I won't stop.

I just love berry-picking.

It just makes me feel at peace. I just love it. I just go where the berries are and just pick. And when my bucket is full, I just go on home.

— Rutie Dicker
From "Berry Picking with Rutie," *Inuit Nunangat Taimannganit*. Used with permission.

Berries of Labrador

CONTENTS

Author's preface 1
How Entries for Each Berry are Structured 3
Notes on Quotations 3

THE BERRIES

- Squashberry 6
- Wild calla 8
- Bristly sarsaparilla 10
- Canada mayflower 12
- Three-leaved false Solomon's seal 14
  - Starry false Solomon's seal 15
- Northern honeysuckle 16
- Crackerberry 17
  - Swedish crackerberry 18
  - Lepage's crackerberry 19
- Red-osier dogwood 20
- Ground juniper 22
- Blackberry 24
  - Purple crowberry 27
  - Pink crowberry 27
- Foxberry 28
- Bearberry 30
- Maidenhair berry 32
- Marshberry 34
- Redberry 36
- *Blueberries & Bilberries* 39
  - Blueberry 40
  - Northern blueberry 42
  - Ground hurts 43
  - Sugar hurts 45

Currant 47	Pear 61
Swampy red currant 49	Wild cherry 63
Bristly black currant 50	Wild strawberry..................... 65
Introduced black currant 51	Bakeapple 67
Twisted-stalk 52	Plumboy 71
Rose twisted-stalk 53	Wild raspberry....................... 72
Bluebead lily 54	Hairy dwarf raspberry............... 74
Red baneberry 56	Poisonberry........................... 76
Dogberry 58	

Illustrated Botanical Glossary 79
Fruit Table 83
Acknowledgements 89
Mixed Berry Jam 91
Bibliography 93
Notes on Contributors 99
Index 101

COMMUNITY & SITE INFORMATION

North
1. Hebron
2. Nain
3. Voisey's Bay
4. Davis Inlet (Natuashish)
5. Hopedale
6. Makkovik/Ironbound Island
7. Cape Harrison
8. Rigolet

West
9. Churchill Falls

Lake Melville
10. Mud Lake
11. Happy Valley-Goose Bay
12. North West River
13. Sheshatshiu

South
14. Paradise River
15. Cartwright
16. Indian Tickle
17. Spotted Island
18. Black Tickle
19. Port Hope Simpson
20. William's Harbour
21. Francis Harbour
22. St. Lewis
23. Mary's Harbour
24. Cape Charles

Straits
25. Red Bay
26. Pinware
27. L'Anse-au-Loup
28. Forteau

AUTHOR'S PREFACE

Berries grow in every part of Labrador — bogs, barrens, forests, gulches, roadsides, yards, mountains, open sand, and wet, shady streambanks. Ripening red, orange, blue, purple, black, or white. Tasting bitter, juicy, sour, sweet, dry. Like wintergreen. Like tobacco. Like no taste. Like, "don't taste, it might be poisonous!"

So here it is, a collection of over forty Labrador berries — some with an important part in Labrador history, some hidden and hardly known. Some with many names, some with only one name. Berries for food, berries for medicine. Berries for birds and animals, but not for people. Berries for everyone. Berries with a single habitat, berries that will grow anywhere. Some for beauty, some for play. Each berry with its own distinct character of root, leaf and stem, fruit and flower.

Labrador voices, describing their experiences, names, and uses for the berries enliven the text, bringing the berries to life.

We will meet the berry whose flower looks like a bird and the berry whose name means "trouble." We will read about the berry with a view, and the one whose roots steal food and drink from other plants. We will know "petal snow," learn of the "little plump prince" of Labrador history, and honour the "queen of berries."

In writing this book, I have kept two audiences in mind: the people of Labrador and readers outside Labrador, botanists and non-botanists alike. The challenge has been to respect these audiences without causing confusion.

Because the book's primary audience is the people of the land, I work to use their vocabulary in the naming and description of the berry species without compromising science. Many botanical terms are explained in the text, others in the glossary.

The berry pieces are arranged by family. Plants in the same family have basic things in common. For example, flowers in the Rose Family have five petals and are usually fragrant. The pear, wild cherry, bakeapple, wild strawberry, and four more Labrador berries are members of the Rose Family.

More than forty berry species in twelve families are put together in this book. A literary mixed-berry jam. Store it on the bookshelf. Open it in winter to have a good read. With this little book, we can have our berries all year long.

— Ellen Bryan Obed

HOW ENTRIES FOR EACH BERRY ARE STRUCTURED

COMMON NAME IN LABRADOR

Subtitle or excerpt from text

L other common name(s) in Labrador   **NA** common name(s) in North American field guides
IT common names in Inuttitut   **IA** common names in Innu-aimun   **S** *Scientific name for the species*

Scientific family name | Family name in English   p #

A NOTE ON QUOTATIONS:

The quotations in this book are the voices of the Labrador people. They represent the residents of communities from Nain in the north to Forteau in the south. They were recorded by the author in personal interviews and by Eva Luther through her many contacts and experiences growing up on the Labrador Coast. And one quote, the epigraph, is transcribed from the video "Berry Picking with Rutie."

There are also journal, book, letter, and oral quotes. These represent a long period of Labrador history from as early as the observations of British naturalist Joseph Banks (1766) and the journals of British naturalist and adventurer Captain George Cartwright (Cartwright, Labrador, 1770–1786), to quotes from 2024, the present day.

The Author Notes reflect the highlights of the author's Labrador berry experiences from 1965 to 1992.

In addition, all poetry throughout the book is written by the author.

THE BERRIES

Squashberry

"It grows up in under the hills. We make jam and jelly. Throw stones away. Put in linen or strainer; squat them out."
— *Anonymous, Author's Notes, Red Bay, 1984*

"We always look for squashberries near water. Sometimes when you are hunting partridge, you smell the berries before you see them. (They have a very distinctive smell with the emphasis on 'stink.') Sure enough, there is a little brook nearby and it's often where you find the French hen/ruffed grouse because they like it near running water. I like the jam with the seeds in it, like chewing the seeds, so wait until they're very ripe and juicy."
— *Liz Dawson, North West River, 2024*

North West River, September

"While in Hebron, I was once offered a bowl full of squashberries from deep inside Hebron or Saglek Bay."
— *Frieda Hettasch, Hebron (1945–59), 1984*

Adoxaceae | Elderberry Family

Red Bay, July

SQUASHBERRY

Flowers for moose, berries for bears

IA mushumin   **S** *Viburnum edule*

> "Squashberries smell like wet, woolly socks while cooking and takes quite a lot of sugar to take the bitterness away. They are better picked before they turn really red and the jelly tastes better too."
> — *Geraldine Curl, St. Lewis, 2023*

Red, round, and juicy, the squashberries are ready to pick. We find them on hardy shrubs that may grow up to two metres tall or on small straggly bushes — by shady brooks, damp places at the edge of the woods, in gulches, or on rocky hillsides. The berries hang in clusters from opposite stems of the bush and are easy to pick. We try one. It is juicy and tart. *Crunch!* Yes, it has a seed inside: one large, flat seed called a stone.

We fill our buckets with squashberries because when we get home, we will:

Squat the berries. Strain the berries.
Boil the juice with sugar.
Boil until there's jelly
to bottle for winter bread.

Black bears look for squashberries too, in damp thickets and on the shores of quiet bays. The berries provide them tasty meals before winter hibernation. But moose prefer to browse the leaves and tender stems in early summer. They even eat the flower clusters. They are not thinking about winter or squashberry jelly.

The squashberry is an attractive shrub. Its pretty milk-white flowers unfurl from dark red buds. Its maple-like leaves grow opposite each other on the stems and turn a deep purple-red in autumn.

We may find the squashberry in almost every part of Labrador, from the Straits area in the south to Saglek Bay in the north.

Adoxaceae | Elderberry Family

WILD CALLA

Most of my family live in the rainforest

NA water arum, water-dragon   **IA** utshishteshu   **S** *Calla palustris*

When we find one wild calla, we find many wild callas! These plants grow together in the shallow, cold water of roadside bogs and on the edges of quiet ponds. They like the sun and they like the wet.

They are connected to each other by thick, spongy rhizomes that travel through the mud. The leaf and flower stalks sprout up separately from these underground stems. The leaves are large, heart-shaped, and curled inward. They turn yellow in the fall.

The tiny, greenish-white flowers have no petals and no sepals. They grow tightly together in a flower head called a spadix. Half surrounding the spadix is the spathe — a leaf-like hood, creamy-white on the upper surface and green beneath. The flowers of the spadix mature into scarlet-red berries with a few oblong seeds inside.

Most parts of the plant are poisonous. They cause an immediate sharp, burning sensation when put in the mouth.

No other flower native to Labrador looks like the wild calla, with its spadix and spathe or "head and hood," but our wild calla belongs to a family of over three thousand species!

Some of the wild calla's relatives are well-known in other provinces — especially the Jack-in-the-pulpit with its purple and white striped hood and the skunk cabbage of many wet habitats. However, most grow in tropical rainforests around the world.

Our wild calla is the only member of the Arum Family in Labrador — found from the Straits area north to Happy Valley-Goose Bay and inland to Labrador West. But its range extends around the world, to Siberia, northern Europe, and Alaska — a tropical beauty in the cold, wet places of the North.

> **Note:** The calla lily commonly sold in garden and flower shops is not a lily at all, but a member of the Arum Family.

Araceae | Arum Family

Wild calla

Happy Valley-Goose Bay

Araceae | Arum Family

Bristly sarsaparilla

Happy Valley-Goose Bay

Araliaceae | Ginseng Family

BRISTLY SARSAPARILLA

Just call me Bristly for short

 uapushimin  *Aralia hispida*

The bristly sarsaparilla grows in abundance in the Lake Melville area, but has it always been here? Is it an introduced species coming with lumber operations or the building of roads and airstrips? Or is it a native species that spread when forests were cut, expanding its sandy niche?

These are questions that botanists often ask about a plant. To find the answer, they look at plant lists that were made in the area in the past. For bristly sarsaparilla, botanists see that its name does not appear on a list of Labrador plants compiled in 1895 by the Canadian Geological Society. But on plant lists compiled after 1939* — there it is!

Introduced or native? We may never know, but we do know that bristly sarsaparilla likes it here. We may find it growing one metre tall. Its amazing root spreads horizontally under the soil for several metres. Its firm, reddish-brown stalk is topped with several green-white flower clusters. The leaves are sharp-toothed and doubly compound.

And bristles? Yes. We could call this plant "bristly" for short. There are soft, fine bristles underneath the leaves, slender bristles on its stems, and tough, woody bristles near the base where stems and root meet.

When late summer comes, "bristly" looks quite fine with its purple leaves and its clusters of dark purple berries. The berries look like blueberries but are inedible and known to make one sick. (The bristly sarsaparilla and blueberry species are not even closely related!)

But songbirds, partridge, foxes, and black bears eat the berries, and moose may browse the tender leaves in spring. How long has "bristly" been growing in Labrador? Wildlife may know, but they do not make lists!

The present known range of bristly sarsaparilla in Labrador is from the Straits area north to Lake Melville area, Central Labrador, and Labrador West.

---

* "*Aralia hispida* — Sandy Banks, Muskrat F. (Doutt 1939, see Abbe 1955). Goose Bay (very common in 1950 and 1952), Happy V., North West R. (1963), common along new road to Muskrat F. in 1967. Native to area?" — Ilmari Hustich, "The Introduced Flora Element In Central Quebec-Labrador Peninsula," 438.

CANADA MAYFLOWER
A starry-white candle on a stem

L & **NA** wild lily-of-the-valley   **IA** anikutshashimin   **S** *Maianthemum canadense*

The Canada mayflower is a lovely little plant with an upright, many-flowered, starry terminal cluster. It is of sweet fragrance, with clasping heart-shaped leaves and a slight zig-zag in its stem. It likes a variety of habitats — shady damp places, open woods, and sandy clearings. The flower often blooms here and there with other woodland plants; it also grows up as a single, shiny leaf in decaying leaf litter. Most lovely of all is when it appears in patches which look like starry-white candles placed tightly together. These patches sprout from networks of thread-like rhizomes in the soil.

Canada mayflower's floral parts grow in fours: four tepals and four stamens.

The one or two-seeded berries are at first yellow-white and spotted brown but when fully ripe, they are ruby-red. Persisting on their stalks all winter, they are food for partridges, Canada Jays, and mice. Not for people. The leaves are foraged by snowshoe hares, mice, and voles.

The Canada mayflower is often called wild lily-of-the-valley. To the casual observer, it does resemble the lily-of-the-valley (*Convallaria majalis*), which has been introduced into North America from Europe and planted in shady habitats throughout the continent. This introduced lily has bell-shaped flowers and is a much larger plant than the Canada mayflower.

Our Canada mayflower is its own starry-flowered self, lighting up every place we find it — from the Straits area north to Ironbound Island off the shores of Makkovik, and as far north as Nain.

Asparagaceae | Asparagus Family

Canada mayflower

Port Hope Simpson

Asparagaceae | Asparagus Family

THREE-LEAVED FALSE SOLOMON'S SEAL

Leaning on the mosses

S *Maianthemum trifolium* (syn. *Smilacina trifolia*)

Fox Harbour

This is a flower of soft, mossy peat bogs. It grows upright on a sturdy stem, but appears weak when we see it leaning on the mosses and other plants of the bog. Its species name is *trifolium* which means "three-leaved." However, there are often two or four leaves on its stem.

It is a small, simple plant with a long, complex name. "Solomon's seal" refers to a closely-related plant not present in Labrador. This plant has thick, knotty rhizomes well-known for their use as a medicine. It is said that long ago a famous king named Solomon put his stamp or "seal of approval" on the healing value of its roots. And some sources say that King Solomon's stamp or "seal" looked like the flat, round scars on the plant's roots!

> **Note:** The flowers and berries of both false Solomon's seals grow in terminal clusters. They appear at the end of their stems.

"If I don't know the name, I'll make one up."
— *George Rich, Rigolet, 1984*

Three-leaved false Solomon's seal

Fox Harbour

The rhizomes of our three-leaved false Solomon's seal are slender and do not have knotty scars. They travel underneath the soil sending up both single leaves and flowering plants. The leaves are narrow, sharp-pointed, and sheathe the stem.

Each flowering plant has three to nine starry-shaped blossoms in a terminal cluster. These tiny white blossoms have six tepals and six stamens. They mature into berries that are first green with reddish spots, then ripen bright or dark red. The berries are inedible and should not be eaten.

Three-leaved false Solomon's seal *is* a rather long common name for a little plant. In fact, it doesn't seem to fit. Could we also call it by another common name? A simple name for a starry-flowered plant of the mossy bogs of Labrador and other northern places around the world.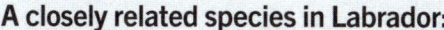

A closely related species in Labrador:

Starry false Solomon's seal (*Maianthemum stellatum* [syn. *Smilacina stellatum*]) is found on the gravelly seaside and cliffs of the Straits area. Sprouting from a spreading rhizome, the plant may reach forty centimetres high. Its stem has many thick, alternate leaves with a starry oblong terminal cluster. The berries of this species are at first green and speckled or streaked with black. They ripen to dark red, nearly black, and are inedible.

Asparagaceae | Asparagus Family

NORTHERN HONEYSUCKLE

The two-eyed berry

The northern honeysuckle is a small, woody plant no more than one metre high. It grows in wet, peaty places along streams and lakesides where it often hides among other shrubs. It is also found on barrens and rocky slopes of mountains. Its oval-shaped leaves grow opposite on the stems. They are covered with minute hairs and often show a tiny point at the end of their leaves. In the springtime, the first leaves and shoots appear purplish green.

Makkovik, August

The northern honeysuckle's funnel-shaped flowers bloom in fragrant, pale-yellow pairs. When we pick a blossom and suck the nectar out, we can say, "This is sweet food for bees and butterflies."

Each pair of flowers joins together to become one bluish-black "two-eyed berry" with many tiny seeds inside. We find the berry juicy with a hint of blueberry-taste. The northern honeysuckle grows north into the Torngat Mountains. Its range extends south and west into the United States and west across Canada.

Because this plant is little-known in most parts of Labrador, we must go on "Honeysuckle Hunts" to find it. We look for its purplish leaves in springtime, the taste of the flower's sweet nectar in summer, and a try of its juicy berries just before fall.

Caprifoliaceae | Honeysuckle Family

CRACKERBERRY
The one that prickles you when you rub the leaves

L crackers, cracker-jack   **NA** bunchberry, creeping dogwood
IT singalâk, kakillânik   **IA** shashakuminan   **S** *Cornus canadensis*

In an old stump? On a hill in the bay? Near the path by our house? We find crackerberries just about anywhere, but their favourite habitat is the boreal forest in cool mossy places under spruce and fir. There they spread in green and creamy-white patches by rhizomes creeping along underneath the soil.

The crackerberry's four white petals are actually bracts: leaves that look like petals. They surround a cluster of tiny, greenish-white flowers that mature into a "bunch" of round, one-seeded red berries. Two or three pairs of leaves make a whorl just underneath the berries; a pair of smaller leaves may appear mid-stem.

The crackerberry *is* edible. Though a fluffy-soft white inside and almost tasteless, it is high in pectin and has been used on the Coast to mix with other berries to thicken puddings and jams. (Hence, the name "pudding-berry" found in some sources.*) And it has even been gathered for crackerberry pie!

Nain, July

> "No one eats 'em but kids."
> — *Bessie Flynn, Forteau, 1984*

> "We call 'em cracker-jack."
> — *John Michelin, North West River, 1972*

---

* *Gray's Manual of Botany*, 8th edition, 1106.

Cornaceae | Dogwood Family

Crackerberry

The berry is a favourite with children for play:

Crackerberry, crackers, cracker, cracker-jack;
Bite a crackerberry hard to hear a crackerberry crack!

In Inuttitut, we say kakillânik — "the one that prickles you when you rub the leaves."

Rub the leaves
together in your hands;
Put your hands,
put your hands
on your face—
it's prickly!

"Cook them; make jam."
— **Kitora Boase, Hopedale, 1984**

Nain, September

A CLOSELY RELATED SPECIES IN LABRADOR:

The Swedish crackerberry, *Cornus suecica*, is closely related to the crackerberry. The plant is smaller and its three to six pairs of leaves are spaced along the stem. Its tiny flowers are dark purple, almost black. Its red berries are edible and slightly sweet. These berries, like those of the crackerberry, have been picked on the Coast to cook with other berries in jam.

This little plant prefers northern peaty and coastal areas. It is often found huddled together in the cracks of rocks on coastal cliffs and hillsides of Labrador and Newfoundland. Its range within Canada extends throughout the Atlantic Provinces, Quebec, and British Columbia in the west.

Its names — Swedish crackers, Swedish bunchberry, and Lapland cornel — give away its widespread presence in Scandinavia. And northern dwarf cornel tells of its place in other northern places around the world — Alaska, Greenland, Iceland, northern Europe, and Asia.

Crackerberry

The crackerberry is also prized for its beauty — creamy-white flowers in summer and dark purple leaves and red berries in fall and winter. It is picked for centerpieces and featured in Labrador art and design.

> "They decorate parkas and tablecloths and they decorate the land."
> — *Clarice Hopkins, Cartwright and Indian Tickle, 1984*

> "They are fluffy inside. Pick for hens. Chew to hear them crack. Rub leaves in hands — prickly."
> — *Anonymous, Author's Notes, Cartwright, 1984*

The crackerberry's range extends from southern Greenland across Canada to Alaska and northeast Asia. In the United States, it grows as far south as New Mexico in the west and Virginia in the East.

> "In one of my recipe books, there's a recipe for crackerberry pie, though I've never made it. It's a very old United Church recipe book made in Goose Bay years ago. Both covers have long disappeared."
> — *Leona Saunders, Cartwright, 2024*

Note: When the ranges of the crackerberry and Swedish crackers overlap, they will often hybridize, thus introducing a new species to the landscape. This species has traits of both plants, making it clearly identified as a hybrid. And that's why it has its own name, Lepage's crackerberry (Lepage's bunchberry in North American field guides), *Cornus ×lepagei*. (×=hybrid)

Hopedale, August

Cornaceae | Dogwood Family

RED-OSIER DOGWOOD

Its branches make "daughter plants"

NA red-twig dogwood **IA** gikuapemak **S** *Cornus sericea* (syn. *Cornus stolonifera*)

The red-osier dogwood is a shrub of river and stream banks, lakesides, and low wet places. It may reach a height of over two metres. When growing in a thicket with alders and willows, it is easily over-looked, but when winter comes, its stems turn red and are easily identified.

Leaves grow opposite on its twigs and twigs grow opposite on its long, thin stems. Flowers appear in flat-topped terminal clusters. Each flower has four white, petal-like bracts and four tiny, green sepals.

In the fall, the berry ripens to a dull, often bluish-white. It has a two-seeded stone inside. The berry is bitter and inedible to most people, but not to several Indigenous Peoples of western North America. Traditionally, they gathered it to mix with the serviceberry, making a "sweet and sour" dish.* Birds nest and mammals hide in the densely-growing red-osier shrubs. When the berries ripen,

---

* From Daniel E. Moreman, *Native American Ethnobotany*, 179-180.

Red-osier dogwood

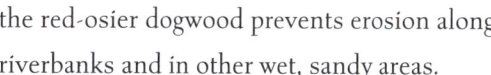

August

black bears and partridge come to feed. So, also, do many other species of birds and mammals.

The red-osier dogwood thicket is a busy place in the fall!

The red-osier dogwood reproduces by seed and vegetatively by stolons. These are the lower branches that lie flat on the ground and put down roots from which "daughter plants" grow. Spreading in this way is called "layering." By "layering," the red-osier dogwood prevents erosion along riverbanks and in other wet, sandy areas.

Because it grows well in many types of soil, it has become a popular plant, sold in nurseries across the country.

A red-osier dogwood thicket makes an excellent nesting place for birds and cover for wildlife. It also provides browse for moose, black bears, beavers, and snowshoe hares.

What is the word "osier" doing in its name? "Osier" tells us that its stems are flexible like those of the osier willow used in basket-making in Europe. (This European osier willow has been introduced into Canada and its stems grown and gathered for basketry here). But the stems of our native red-osier dogwood are harvested for basketry as well. In fact, Indigenous peoples of North America were making cradles, baskets, and snowshoes from its stems long before it was ever known as the red-osier here.

In Southern Labrador, the stems are often woven into bright red-twigged wreaths for winter doors.

GROUND JUNIPER

The berry that's a cone

L low juniper **NA** common juniper **IT** kisittutaujak **IA** kakatshimin **S** *Juniperus communis*

Hopedale, August

The ground juniper is an evergreen shrub whose cones look like berries. The seed cones of the female plants ripen in stages and may take three years to mature. We often see green, blue, bluish black, and whitish-bloomed berries on the same branch.

Male juniper plants have small, light brown flowers with many scales full of pollen. Wind carries this pollen to female plants, whose flowers develop into fleshy cones. Two or three seeds are inside each cone.

The ground juniper is a prickly shrub whose needles have very sharp points. It spreads in large, circular mats on rocky hillsides and sandy places. Birds find a nesting place and small mammals a shelter in the juniper mats. Many species of birds and mammals feed on the berries. And caribou have been seen eating the branches on the barren grounds when other food is scarce.

"Berries are bitter. They grow around coast, in North West River and north on the barrens, about two hundred miles north of Churchill. Porcupines eat the berries."

— *Max Maclean, Churchill Falls and North West River, 1984*

Cupressaceae | Cypress Family

Ground juniper

Many cultures have used parts of the ground juniper for medicine. In Labrador, Inuit, Innu, and settlers have boiled the branches and berries or eaten the berries raw to treat flu, colds, and other conditions.

The berries are aromatic; they have a pleasant smell. In many places outside Labrador, they are used to flavour wild meat, to boil for tea, or to make into jewelry.

The ground juniper grows from the Straits area to the North Coast and inland barrens. It has a wide range both inside and outside Labrador. In fact, it is considered to be the most widespread conifer in the world!

> "Steep the juniper with its berries for cramps. Boil rind; wash eczema with juice."
> — *Dave Dyson, Black Tickle, 1984*

> "Eat berries raw for flu and cold. Boil branches and berries and drink for colds."
> — *Kitora Boase, Hopedale, 1984*

Cupressaceae | Cypress Family

BLACKBERRY

Smoke for trout, dye for hair

L curlewberry **NA** black crowberry **IT** paungak **IA** assimin **S** *Empetrum nigrum*

Red Bay, July

The blackberry is a plant with many stories to tell. It remembers the huge flocks of curlew* descending upon its coastal mats in their fall migrations to fill themselves with berries. It remembers codfishing days and blackberry tea biscuits. It recalls its branches smouldering in smokepots to keep flies away from the fishing stages.

The blackberry speaks of its purple juice boiled into a summer drink or fermented to become a Christmas wine. It knows it has been used to dye coastal grasses for weaving into beautiful baskets. And it cannot forget when a long time ago, a herring factory owner on the South Coast actually dyed his hair with blackberry juice.

For generations, blackberry branches and sod have been used in smokehouses to give a tasty flavour to trout, salmon, and char. Its branches were often kept through the winter to

"We use bush and sod for smoking trout and salmon."
— Sam Learning, Paradise River, 1984

---

* Historically, this species was called the "Eskimo Curlew." It is now extinct. See Charles W. Townsend's quote, next page.

Ericaceae | Heath Family

be boiled and steeped for tea. Their fragrance brought the salty summer smell of blackberry mats by the sea.

The blackberry is watery, sweet, and a bit crunchy, with several tiny nutlets inside. It's what the northern sun and soil and sea mist have put together: hard to describe, easy to taste. We pick enough to save for winter puddings, cakes, and mixed-berry jam. But we always eat a few on the way home.

On the North Coast, Inuit combine blackberries in a pot with fried cod livers (siva) for a dish called sivalik. The berries can also be mixed with cut-up char roe and a drop of seal oil to make sivalik.

Even the blackberry mats have something to say. Their woody branches and small (three to seven millimetre) needle-like evergreen leaves are a home for mice and a place for birds to nest. Arctic hare feed on the branchlets. And when their berries are ripe, many animals and birds come to forage: black bears and polar bears, gulls, ravens, and partridges. Migratory birds flying south in the fall depend on the berries for food. When Canada Geese feed on the blackberry, their meat is sweet and stained with blue.

The blackberry's flower can boast that it is one of the first Labrador plants to bloom in spring. As soon as the snow melts from the mats, the tiny purple flowers appear. They are so small (two millimetres across) that we must look closely to see them.

Though blackberries grow in forested and alpine areas inland, they are most abundant near the Coast. And there in the salt spray they could say,

> "We also made wine from the blackberry. It was our Christmas wine."
> — *May Pardy, Cartwright, 1984*

> "The birds were delicious eating. They fattened almost to bursting on the Empetrum, or curlewberry, so abundant along the coast. The fishermen kept their guns loaded, and shot into the great flocks as they wheeled by, bringing down many a fat bird. About 1888 or 1890 the curlew rapidly diminished in numbers, and at the present perhaps a dozen or two, or possibly none at all, are seen in season."
> — *Charles W. Townsend, quoted in* **Labrador: The Country and Its People**

Blackberry

"Here we ripen to be the largest, juiciest, tastiest blackberries of all."

The blackberry is a circumpolar, circumboreal species with the most widespread range of any member of the heath family in the world. Its generic scientific name, *Empetrum*, tells us why.

It comes from the Greek *en petros*: "on the rocks." The plant is able to spread itself over the rocks in the harshest of northern climates and on alpine summits and coastal headlands to the south.

Its scientific species name *nigrum* means "black," hence one of its English names, "crowberry," and one of its Russian names, voronika, "little raven." And in Lapland, kaarnikka, simply "raven."

"Old Alfred Spearing says it was Mr. Bannister from England, owner of the herring factories in Trap Cove and Cape Charles, who dyed his hair. And that he picked the blackberries for him!"
— *Eva Luther, St. Lewis, 2020*

"When we moved out from the salmon places in the bay to the Coast for cod-fishing, we'd put small children in a patch of blackberries. In no time they would be like little puddings. We made tea biscuits piled with blackberries and just enough flour to stick 'em together."
— *Clarice Hopkins, Cartwright and Indian Tickle, 1984*

BLACKBERRY HAIR

There once was a man down the bay
who wanted black hair, not grey;
so he decided to use
some blackberry juice
but his hair became purple that day.

Now this gentleman man was so vain
that no one dared mention the stain
that ran down his cheeks
in purplish streaks
whenever it started to rain.

Two closely related species in Labrador Straits area:

Pink crowberries (*Empetrum eamesii*), also known as rockberries, grow in dense, ground-hugging mats. Their young branches are densely covered with white, woolly hairs. Their oblong, leathery leaves are very small, only four to five millimetres long and oval in cross-section rather than flat. Their berries ripen from green to pink to light red and are three to five millimetres across. They are edible and juicy, but with not much flavour.

Pink crowberries grow in dry, exposed, rocky, or alpine habitats in southeastern Labrador.

Purple crowberries (*Empetrum atropurpureum*) grow in loose mats. Their branches are commonly seen trailing out from their mats over the rocks and sandy soil. The young branches have a thin covering of white, woolly hairs. The oblong, leathery leaves are four to seven millimetres long.

Their berries ripen from dark red to deep reddish purple and are five to nine millimetres across. They are also edible and juicy, but with not much flavour.

Purple crowberries grow on coastal headlands and in alpine habitats in southeastern Labrador.

> "To protect its leaves from cold, drying winds, the hairy margins of *Empetrum* leaves curve under and meet at the bottom of the leaf, enclosing the lower surface inside, and forming a white line that resembles the zipper of a jacket."
> — Sue Meades (botanist), 2024

FOXBERRY

We see the hills burning with colour

 cobbler   alpine bearberry   kallak   S *Arctous alpina*

Red Bay

The foxberry grows in large mats that spread every which way, covering coastal hills and inland alpine summits. Its shreddy-barked branches are reddish-brown. Its wrinkled, leathery leaves are deciduous. They become scarlet, then burgundy after the first frost. Foxberry leaves make the autumn landscape beautiful. When we travel by speedboat or by plane, we see the hills burning with colour.

"When we lived on Spotted Island, we used to call this berry a 'cobbler.' My mother told me it was poisonous, so we just used it for play. Years later I found out it was safe to eat, but I still won't eat them because my mother told me not to."

— *Eva Luther, St. Lewis, 2020*

Ericaceae | Heath Family

Foxberry flowers open with the new leaves in the spring. They are urn-shaped and greenish-white. We find their little clusters amidst whitened leaves and withered berries from the year before. The new berries are plump red, and ripen purplish-black. Each berry has five tiny nutlets inside. Even in the fall at berry-picking time, we can still see the remains of last year's leaves and berries. The dead leaves have protected the plant from wind and cold.

We pick several kinds of berries on the hills — redberry, blackberry, ground hurts — but not the foxberry. Even though it's edible, it has a very unpleasant taste. We don't eat it!

"Foxberry, bearberry" — these names tell us that red foxes and black bears eat the berries. They forage in late summer and fall, while mice, vole, and lemming feed on them in wintertime. These little mammals are always looking for berries as they scurry about making trails underneath the snow.

> "Toward the end of summer (Nain, 1940) there were several species of berries to harvest, preserve and store for winter. One time, wandering along the shore, we came across a large patch of red berries. I took them for partridgeberries — a form of cranberry — and assured Doris they were both safe and delicious. We picked enough for a pie, and ate the pie but found it tasteless... we had eaten bearberries [foxberries], alleged to be indigestible and even poisonous. I can report that the only damage in this case was to my ego."
>
> — *F.W. Peacock, from* Reflections from a Snowhouse, *67.*

BEARBERRY

The bear's grape

IT kinnikinnick   **IA** atikminan   **S** *Arctostaphylos uva-ursi*

The bearberry is not a common berry in Labrador, but it is here.

Look in sandy places in Lake Melville area, Labrador South, and the Straits area.

Look for a plant that can run its thin, flexible branches over the ground for several metres.

Look for thick mats of leathery, paddle-shaped leaves.

Look for pink or white urn-shaped flowers at the end of the branches in summer.

Look for round, red berries in the fall or winter. Taste one and you will quickly discover that the berry is dry, mealy, and insipid. It has no flavour. The stone inside the berry is made up of several nutlets fused together.

Look for a plant that would be perfect for winter decorations — if it were plentiful enough. The bearberry is sold in plant nurseries across North America for a beautiful ground cover and for the prevention of erosion in sandy soils.

The bearberry is a widespread species in Alaska and the Canadian and American West. In these regions it is commonly called kinnikinnick. Indigenous peoples and early settlers dried its leaves for a tobacco substitute and used it as medicine (the word kinnikinnick is from Ojibwe; it originally meant any herbal smoking mixture). The Inupiat of Northern Alaska use the bearberry in their akutuq, a combination of animal fat, seal oil, berries, and snow.

Bearberry

Many animal and bird species eat the berry in the West, but grizzlies and black bears are its chief foragers. In fact, both its common name, bearberry, and its scientific names, *Arctostaphylos* and *uva-ursi*, were inspired by bears. *Arctostaphylos* means "bear and bunch of grapes" in Greek. *Uva-ursi* means "the bear's grape" in Latin. Black bears might be a chief forager in Labrador, too, if the berry were as plentiful here as it is in the West!

The bearberry is a circumpolar species, growing in Greenland, Iceland, Northern Europe, and Asia. Here in North America, it grows from Labrador south to the Pine Barrens of New Jersey and the mountains of Virginia. Its range extends across Northern Canada to Alaska and appears south on the Coast of California and the length of the Rocky Mountains to New Mexico.

Note: The bearberry and the alpine bearberry (foxberry) are closely related, but have distinct differences.

The bearberry leaves are green all winter. Its berries are dry and mealy. Its favourite place to grow is in the sand.

The foxberry leaves wither and die in the fall. Its berries are plump and juicy. Its favourite places to grow are on coastal barrens and inland mountain summits.

Mud Lake, October

Ericaceae | Heath Family

MAIDENHAIR BERRY

A tiny taste of white

L wintergreen, mint, minty berry, magna-tea, capillaire   **NA** creeping snowberry
IA pineminanish   **S** *Gaultheria hispidula*

The thin, slightly woody stems of the maidenhair berry make trails in the cool damp mosses of boreal forests. They are found throughout Labrador from the Straits north to the Nain area. They bear tiny evergreen leaves and tiny bell-shaped flowers that become tiny white berries. The entire plant is flecked with rusty-brown hairs.

The berry is actually a many-seeded capsule (a type of dry fruit) enclosed by the flower's sepals. As it grows, the sepals enlarge, becoming fleshy and white until the capsule is a berry-like fruit — a fruit not much bigger than an ant's egg.

The maidenhair berry is a favourite plant for its wintergreen fragrance and flavour. For generations, its leaves and berries have been boiled as "magna-tea" — a fragrant substitute for black tea. And it is always a tasty, refreshing nibble on a walk.

Wildlife prize it, as well. Partridges are said to get fat quick on its berries. And deer mice may fill their cheeks as they scurry along the forest floor looking for food.

This is a plant of many names. "Maidenhair berry" originates in Newfoundland. "Maidenhair" refers to its dainty leaves that look like those of the maidenhair fern, which grows in the western part of the Island. Capillaire is French for capillary, meaning hair-like. Then there are Labrador local names such as minty berry and toothpaste berry, magna-tea and wintergreen. Almost every community gives a name of its own.

North West River, September

Ericaceae | Heath Family

Maidenhair berry

Forteau

"We drink it in spring when we are out of tea."
— *Muriel Andersen, Makkovik, 1984*

Outside Labrador, it is commonly known as creeping snowberry — the little trailing vine of the forest and mossy bog that hides its berries under its leaves.

Exploring the spruce-fir forest and edges of bogs, we look for the maidenhair berry. And when we see it,

we pull abroad the mosses
for berries
out-of-sight,
to taste
the taste of wintergreen
in a tiny taste
of white.

"My late husband, Brian Michelin, told me that his mother, Louisa Michelin, used these berries. She picked them and dried them in a cloth bag; then she steeped a tea to treat family members when they had a fever. He told this to me circa 1978."
— *Liz Dawson, North West River, 2024*

Ericaceae | Heath Family

MARSHBERRY

Speckled eggs in a nest of mosses

NA small cranberry, bog cranberry, wren's egg cranberry **IA** massekuminan
S *Vaccinium oxycoccos*

Happy Valley-Goose Bay, September

With its thin evergreen vine, the marshberry travels over the mossy bogs of Labrador — all the way to northernmost Labrador-Ungava. Its wiry, woody branches are so thin and its tiny leaves are so scattered that the plant is easily overlooked until its plump berries appear. We walk to find them in autumn, in the tang and tartness of a wet, boggy place.

The marshberry flower is small, but striking. Because it looks like the head of a crane, the plant was given the common name "crane berry" (now "cranberry"). The flower's four pink petals bend backwards to look like the shorebird's head; its eight stamens press against the style to form its beak. The flower rises from a very thin stalk that resembles the crane's long neck. A pair of tiny bracts appear halfway up its flower stalk.

When marshberries are ripening, they look like speckled birds' eggs in a nest of mosses. With the frost, the berries turn dark red. They are edible, but sour. They must incubate under the wings of winter snow to perfect their sweetness. Then, when the snow melts off the mosses, they are soft, juicy, and sweet — ripe for the

> "I remember the marshberries especially in spring, picking them and Aunt Susie Rich making those delicious pies with marshberry jam."
> — *Sarah Baikie, Rigolet, 2020*

Marshberry

> "They are best to pick in spring after the snow."
>
> — *Elizabeth Goudie, Happy Valley, 1972*

picking! But we must gather spring marshberries with care. We don't want to break their skins and stain our fingers with juice that was meant for jam or the first berry pie of the year.

Outside Newfoundland and Labrador, the marshberry is called the small cranberry because it is closely related to the large cranberry we find in stores. These commercial cranberries (*Vaccinium macrocarpon*) grow from Newfoundland to Ontario, and in British Columbia. They are also found in Alaska and in many states south of the border. They are raised in giant bogs for export to world markets.

Though our small cranberry has a limited harvest, it has a far broader range than its larger cousin. It is a subarctic and circumboreal berry, growing in cold places all around the world.

Pinware, July

Happy Valley-Goose Bay, September

Ericaceae | Heath Family

REDBERRY

Wait until after the first frost

L partridgeberry **NA** mountain cranberry, lingonberry **IT** kimminak, kimminait **IA** uishatshimin **S** *Vaccinium vitis-idaea*

There are many red berries in Labrador, but there is only one redberry. Growing on coastal barrens, in open lichen woods, sandy places, bogs, and cracks in rocks, the redberry is almost everywhere. Tart, tasty, and full of vitamins, easy to pick and easy to store. It is Labrador's most important land crop.

> "I've seen the snowbirds' breasts all pink from eating redberries."
> — *Aunt Flo Baikie, North West River, 1972*

The redberry is a beautiful little plant with its shiny leaves and clusters of pink, bell-shaped flowers. Its rhizomes creep along just under the ground. Slightly woody stems sprout up from the rhizomes to form large evergreen mats. In the fall, these mats are full of berries — round, red gems ready to pick. But we must wait until after the first frost. That's when the fruitworms come out of the berries.

It is September when we see people heading off to their favourite berry-picking spots: some to hills close by, others to far-off barrens, some to hidden spots among the birches. Others do not tell where they go. Many families on the Coast travel in their speedboats to nearby islands. With a gun for ducks, a kettle for boil-up, scarves for flies, and buckets for berries, they hope to return with enough berries for winter.

Red Bay, July

> "They never spoil. They taste the same no matter if they been in the freezer five months or five years."
> — *Roxanne Notley, Port Hope Simpson, 2020*

Ericaceae | Heath Family

Redberry

> "My favourite is picking them in the spring as the snow melts away from the barrens. So sweet and juicy. I also make a drink from the juice because redberries are very high in iron and no side effects like iron pills."
> — *Joyce Lee, Red Bay, 2020*

In years when vegetables and fruits were scarce, redberries provided the nutrition for people to stay healthy through the winter. They were also used as medicine. "Eat them raw for a sore throat," people would say.

In Northern Labrador, redberries are picked just before the snow. They are also harvested in spring when the snow melts from the patches and the berries are sweet. Inuit have long made a popular dish that combines kimminait with seal fat, char roe, and fermented seal oil. They also cook the berries to make jam.

Redberry jam, cakes, puddings, and quick breads are favourites in every Labrador kitchen.

Red Bay, July

> "I like the tartness, love redberries in pie, and my favourite jam! I like the memories of having a barrelful in the winter and cutting them out to make jam."
> — *Madeline Flynn, Forteau, 2020*

Ericaceae | Heath Family

Redberry

Many migratory birds feed on the snow-sweetened berries as they fly north in the spring. Partridge depend on them as a source of food, especially after the snow melts — hence the name "partridgeberry." And black bears, martens, foxes, mice, and voles forage the redberry mats in their own seasons.

This red berry is known by more than forty names around the world. Mountain cranberry, lowbush cranberry, and lingonberry in North America, cowberry and red whortleberry in Scotland, lingon in Sweden, and kokemomo in Japan are just a few. But in Labrador we call it simply "redberry" because it is *the* red berry of our land. The red berry we prize. 🪶

> "I remember when I was young, my mom picked redberries before the snow for medicine. When babies got problem with teeth, all red inside, she used a little squeezed redberries inside the mouth."
> — *Elizabeth Penashue, Sheshatshiu, 2020*

Red Bay, July

BLUEBERRIES & BILBERRIES

Bilberries

Every place in Labrador has its blue berries, and every community has its own names for them. These blue berries can be divided into two groups: blueberries and bilberries.

Both blueberries and bilberries belong to the *Vaccinium* genus, but they have distinct differences:

Bilberry
- The fruit mostly grows singly in the leaf axils.
- The bilberry fruit varies in shape.
- The style of the bilberry flower persists as a thread-like bill coming out of the top of the fruit.

Blueberry
- The fruit grows in a cluster at the ends of its branches.
- The blueberry fruit is round.
- The sepals of the blueberry persist to make a five-pointed crown.

Blue berries, blue berries, everywhere blue berries —

sweet, not sweet, and in between;
dark blue, light blue, whitish bloom,
unripened white or pink or green.

Blue berries, blue berries, everywhere blue berries —

on hard-wind hills by the side of the sea,
in soft, wet moss, in sun dry sand,
in company of shrub and tree.

Blueberries

Blue berries, blue berries, everywhere blue berries —

ground hurts, sweet hurts, tobacco hurts, too.
Bilberries, whortleberries, huggleberries, hurts —
everywhere blue berries and blueberries, blue.

Ericaceae | Heath Family

BLUEBERRY

A five-pointed crown

L tobacco hurts **NA** common lowbush blueberry **IT** kigutanginnak
IA inniminan **S** *Vaccinium angustifolium*

The blueberry wears a five-pointed crown at the top of its fruit. The points of the crown are the flower's persisting sepals. These "crowned blueberries" are a widespread species in Labrador from the Straits area north to Nain. They grow in sandy soils up and down the Coast among redberries and other species of blue berries and in patches of their own. They are especially abundant inland in areas where forests have burned or been cut down during logging operations.

The blueberry plant varies in size from five to fifty centimetres tall and spreads by rhizomes, its underground stems. In springtime, the new leaves are tinged the colour of bronze. The white and pinkish-white flowers bloom in terminal clusters. In August when the berries begin to appear, we often see several stages of ripening in one cluster — white, pink, purple, and blue. The round, fully-ripe blueberries are covered with a whitish bloom. Sweet and tasty, but sometimes not so sweet and tasty.

Red Bay, July

Ericaceae | Heath Family

Blueberry

North West River, September

Soil, organic matter, water, and light affect the blueberry plant's growth, its berry size, and its taste. That's why we could say, "*Sweet and tasty!*" or "*hardly any taste!*" or "*tastes like tobacco*" of the same berry. It all depends on where we find it growing.

Because the blueberry is abundant and easy to access in the Lake Melville area, it is a favourite berry to harvest there. When its leaves are turning red and an early fall wind is chasing flies away, we go to pick blueberries. Jam, cake, and winter pie are on our minds. Black bears, well-known for their love of blueberries, know these places, too. They will eat their litres on the spot, but we will take our litres home.

The blueberry is not only popular in Labrador. It is the famous lowbush blueberry of Newfoundland, Quebec, Nova Scotia, New Brunswick, P.E.I., and the state of Maine. In these places, it is harvested in huge commercial operations and sold to markets in over thirty countries around the world.

NORTHERN BLUEBERRY

They tastes like tobacco

 tobacco hurts  northern blueberry  kigutanginnak   *Vaccinium boreale*

The northern blueberry is similar to the common blueberry in appearance, but with closer consideration, we can see its differences. In height, size of leaf, and flower, the northern blueberry is a smaller species. Its leaves are bright green and hairless. Its rootstalks spread on the surface, not as rhizomes underground.

The northern blueberry is more common on the Coast and extends further north than the common blueberry.

> "Tobacco hurts are with a crown. They tastes like tobacco."
> — *Anonymous, Author's Notes, Port Hope Simpson, 1984*

> "Backy hurts don't have much taste."
> — *Warrick Chubbs, St. Lewis, 1984*

Churchill Falls Road, July

Ericaceae | Heath Family

GROUND HURTS
The berry with a view

 blueberry  alpine bilberry, bog bilberry  pungajok  nissimin *Vaccinium uliginosum*

Red Bay, July

The ground hurts takes us up from the harbour to the barrens where we look for icebergs or watch for the coastal boat to come in. There we find it growing in places of little soil and much wind. Its tough, sturdy branches spread flat across the lichen-covered rocks.

On these windy barrens, ground hurts' branchlets rise only eight to twenty centimetres above the ground. They bear smooth, round, blue-green leaves that narrow to the base. The flowers appear singly in the leaf axils. But sometimes they grow in clusters of two or three at the end of the branchlets. They are bell-shaped and coloured white or pinkish white. They mature into deep blue, almost black berries with a whitish bloom.

Wherever they grow, the berries vary in shape, size, and taste. With their soft skins, they do not store well, so as soon as the berries are picked, they quickly become jam, cake, and pie.

Though most common on the coastal hills, the ground hurts appear in a variety of habitats throughout Labrador. They can be found in shady places and on rocky shores, and in sheltered gulches and bogs where they may grow one metre tall.

Ericaceae | Heath Family

Ground hurts

> **Note:** The style of the ground hurts' flower persists and looks like a thread-like bill coming out of the berry. But the name "bilberry" does not refer to this. "Bilberry" comes from the Danish word bollibar, which means "dark berry."

From its wet habitats, it gets its common name bog bilberry and its scientific name *uliginosum*, "of swampy places." Both in the Arctic and in the mountainous areas south of Labrador, it is known as alpine bilberry.

But on the Labrador Coast, it is ground hurts. This means tough, sturdy branches spreading across the windy barrens, red leaves colouring the barrens in autumn, blue berries speckling them in summer. And ground hurts means the place where we go to look out to islands and bays, icebergs and boats, and down to the harbour where we will take our berries home.

Red Bay, July

"The berries of the *Empetrum Nigrum*, and likewise, some delicious blue berries which grow on a small shrubby plant, called Ground Whortle, both of which are now ripe, are what the curlews delight to feed on. These not only make them uncommonly fat, but also give them a most delicious flavour."

— *Captain Cartwright's journal, August 26, 1770*

SUGAR HURTS
Purple sweetness on my tongue

 sweet hurts, huggleberries  dwarf bilberry  pungajok  *Vaccinium cespitosum*

When we pick blueberries, we often come upon a little bush whose berries are sweeter than all the others. We call out "sugar hurts!" and eat them right on the spot. Never enough for jam or pie — just a brief, sweet taste of blue. They grow here and there among the redberries and blueberries in the sandy soil of summer.

The plant is small, only five to thirty centimetres high. Its branches grow in a dense cluster. The fine-toothed leaves are pointed or round at the tip and narrow towards the base. The flowers are pinkish-white or dark pink. They mature into blue berries covered with a light bloom. To pick them, we lift the tiny leaves and find the berries nodding underneath.

> "We call 'em huggleberries. Raise 'em up and hook 'em under."
> — *George Poole, St. Lewis, 1984*

The sugar hurts is actually a bilberry. It differs from the blueberry in several ways. Its fruit varies slightly in shape, from round to oval. The thread-like style from its flower sometimes stays and sticks out of the top of the berry. The berries grow in the leaf axils, usually singly, but sometimes in groups of two or three.

Churchill Falls Road, July

Ericaceae | Heath Family

Sugar hurts

Sugar hurts is called dwarf bilberry across the subarctic of Canada and Alaska. It also keeps this name where the plant extends southward into the alpine areas of the United States. But in Labrador, we will keep "sugar hurts" and "sweet hurts" and "huggleberries." These are the sweetest names we know.

Note: The word "hurt" comes from the old English word "whortleberry." "Whortleberry" has long been used in Scotland and England for the berries of the *Vaccinium* genus. In these two countries, our redberry is the red whortleberry and our alpine bilberry or ground hurts is the bog or ground whortleberry. When Scottish and English settlers established themselves in Newfoundland and Labrador, they shortened the name "whortleberry" to become whort, hart, hert, hirt, hort, or hurt!

Sugar hurts! Sweet hurts!
I pick the sugar berries
one by one.

Sugar hurts! Sweet hurts!
Leave a purple sweetness
on my tongue.

Sugar hurts! Sweet hurts!
Grow on little bushes
in the sand.

Sugar hurts! Sweet hurts!
Enough to hold the summer
in my hand.

"The huggles were the sweetest of the berries. They would pick them on the Burned Hill and in the Laid at Deep Water Creek. Uncle Sol would take a handful of berries, smack his lips and say, 'They're some good, Florric.'"
— *Warrick Chubbs, St. Lewis, 2020*

CURRANT

Those smelly berries

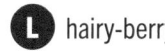 hairy-berry  skunk currant  kimminaujak  minitshimin  *Ribes glandulosum*

The currants in our stores have been picked, dried, and packaged in places far away. They are ready for mixing into any kind of fruitcake or currant bun. But we also find currants here in Labrador, growing ripe and ready to pick — garden and yard currants and three species of wild currants.

The most common of the wild currants is called simply "currant," or in Labrador South, hairy-berry. We find its smooth gray and brownish branches

> "We picked red currants, cooked, strained, and used the juice to make a fine jelly."
> — *Clarice Blake Rudkowski, Happy Valley, 2020*

> "While looking for squash-berries one day, I came across the hairy-berries. They taste much better than squash-berries do and they smell much better, too."
> — *Leona Saunders, Cartwright, 2020*

> "A beautiful berry for mixed-berry jam."
> — *George Poole, St. Lewis, 1984*

straggling over rocky places and clearings. It is found just about everywhere in Labrador, both inland and on the Coast. The plant may grow up to one metre high. Its maple-like leaves come as the first green of spring. We pick its dark red, hairy berries from their small, drooping clusters in late summer.

Both leaves and fruit have gland-tipped hairs which give off an unpleasant odor when crushed, hence the name "skunk currant."* But the smell of the fruit does not spoil the currant's taste, whether eaten fresh or boiled into jelly.

In Labrador, we do not usually dry our currants, but we could. Traditionally, we pick them for a bright red jelly that is finer and more excellent than found in any store.

---

* The plant gets this name from the striped skunk, absent from Newfoundland and Labrador, but widespread in temperate regions across Canada and the United States.

Currant

"The first green leaves of spring."
— *Captain Cartwright's journal, April 17, 1772*

Forteau, July

North West River, September

"My brother and sister-in-law gets lots of currants around Forteau area. They are never without those smelly berries, but makes beautiful jams."
— *Clarissa Smith, Bradore Bay, Quebec, 2020*

Grossulariaceae | Gooseberry Family

SWAMPY RED CURRANT

At home in the spray zone

 Ribes triste

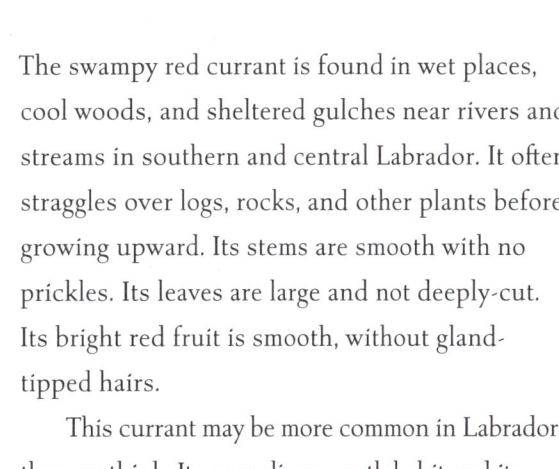

Churchill Falls, September

The swampy red currant is found in wet places, cool woods, and sheltered gulches near rivers and streams in southern and central Labrador. It often straggles over logs, rocks, and other plants before growing upward. Its stems are smooth with no prickles. Its leaves are large and not deeply-cut. Its bright red fruit is smooth, without gland-tipped hairs.

This currant may be more common in Labrador than we think. Its sprawling growth habit and its leaves are similar to those of the well-known currant. And though the swampy red currant berries are smooth and the currant berries are hairy, they are both red. At a distance, the plants and berries may appear to be the same.

"I found the swampy red currant in the spray zone of Churchill Falls, ten years after the falls were stopped. It was enjoying the damp, nutrient-rich habitat with its close relative, the common red currant."

— *Author's Notes, 1984*

Grossulariaceae | Gooseberry Family

BRISTLY BLACK CURRANT
Where mosquitos and black flies like to be

IA kakumin **S** *Ribes lacustre*

Happy Valley-Goose Bay, September

This currant species thrives in cool, damp, shady woods and stream sides — mosquito and blackfly places we avoid in summertime. But the bristly black currant is there, sometimes growing two metres high! Its young branches are covered with prickles, its older ones with flaky bark. Its leaves are deeply cut and smaller than those of the red currant. Its purplish-black fruit is covered with gland-tipped hairs. Crushed, the berries give off a smell; tasted, they are very sweet.

When we do come upon a bristly black currant bush, we may not find enough berries for jelly, nor even a handful. Maybe not more than three or four berries! But that's enough of its amazing taste to last a wintertime. Then when summer comes again, we will look for it in those hidden damp and shady places where mosquitos and blackflies like to be.

> "I came upon this berry once, in 1992, on a small protected, shady island in the Grand River not far upriver from Muskrat Falls. The plant was almost two metres tall. The fruit appeared here and there on drooping stems. So boldly sweet, so plumply delicious I would paddle the river again for it. But the bristly black currant is lost now with its little island in the deep trenches of a monstrous reservoir."
>
> —*Author's Notes, 1992*

Grossulariaceae | Gooseberry Family

Bristly black currant

"When I was a child, my mom always had black currant to combat sore throats and colds.
When I came to Labrador twenty-one years ago, I ordered different varieties of black currant. One fall I picked thirty gallons from my bushes!
My husband, Warrick, tells me that the berries were planted in areas called bawns by the ocean. In these places the soil is enriched from drying fish on rocks and wooden structures called flakes.
We shared our plants with a nurse in Hopedale. She put them in buckets and kept them in her basement. Then she planted them outside. The black currant is now growing as far north as Hopedale.
The leaves make a nice tea. We make a beautiful black currant juice. It is delicious."

— *Elaine Chubbs, St. Lewis, 2024*

"My family tells a story that long ago the black currant bushes were brought over from England in a great aunt's apron and planted at the Cape. As long as I remember, black currants have been growing here."

— *Roy Pye, Cape Charles, 2024*

The Introduced Black Currant

When we speak of the black currant, we generally are not referring to the bristly black currant, but to the cultivated black currant which was introduced into Labrador several generations ago. This currant spreads easily from where it is planted; it becomes a patch or pops up here and there about the landscape. It has been transplanted successfully in many places inland and on the Coast.

Grossulariaceae | Gooseberry Family

TWISTED-STALK

The berry with a twist

 lily   clasping-leaved twisted-stalk **S** *Streptopus amplexifolius*

Twisted-stalk is a tall lily with a forked stem that grows up to one metre high. Its leaves are large and pointed, clasping the stem. As in all lilies, the leaves are parallel-veined.

The twisted-stalk's flowers hang singly or in pairs underneath the leaves. If we look closely, we can see the thread-like twisted stalks that give the plant its name. The flowers have six pale, white tepals and six stamens.

The many-seeded, oval-shaped berries ripen red. They are inedible and considered poisonous.

Twisted-stalk is fun to find. It may appear just about anywhere in Labrador, from the Straits area into the Torngats north of Nain. We may come upon it in the deep woods by a spring in Paradise River. We may find it in a mossy place on Ironbound Island outside Makkovik or behind an old tilt along the Grand River. We may see it growing among grasses and sedges north on the edge of the treeline.

But here's the twist: sometimes we may not notice it. Because the twisted-stalk hides its pale flowers underneath its leaves and often blends in with other plants, it might not be identified until early fall. Then its bold, red fruit will give it away.

Forteau, July

> "We were told if we ate these berries, we would go to sleep and never wake up."
> — Bella Lyall, *Voisey's Bay, 1972*

Liliaceae | Lily Family

Twisted-stalk

> "I never heard a name for them in Nain where they grew by the old water dam."
>
> — *Doris Peacock, Nain (1940–57), 1984*

> "Hiking inside Hopedale in early August, I came upon a drift of snow melting into a brook. And there I found the twisted-stalk; it was blooming in a warm green patch of summer."
>
> — *Author's Notes, 1984*

A closely related species in Labrador:

Rose twisted-stalk (*Streptopus lanceolatus*) is a close relative of the twisted-stalk, found mainly in the Straits area. It grows in cool, moist places. Its bell-shaped flowers are pink or rose-purple. The berries are dark red. The leaves touch the stem without clasping it. They are slightly hairy and their margins are finely-toothed. The whole plant shows a general hairiness.

The British naturalist and explorer Joseph Banks first recorded this pretty rose twisted-stalk in Chateau Bay, Labrador in 1766.

Liliaceae | Lily Family

BLUEBEAD LILY

Like a protected princess

 NA clintonia IA anikamin S *Clintonia borealis*

This is the lily of light-yellow flowers in summer and of blue bead-like berries in fall. It is also the lily of tender shoots in spring that unfurl into large, glossy basal leaves.

We find the plant in cool, moist woods, in quiet patches underneath birches, and in gulches on the coastal tundra. It grows up from rhizomes that spread underneath the soil.

Like most members of the lily family, the bluebead lily's flower parts grow in threes and sixes. Its bell-shaped blossom has six tepals and six stamens.

> "I found the yellow clintonia growing like a protected princess in almost every gulch of the coastal tundra from William's Harbour to Francis Harbour"
> — *Author's Notes, 1984*

Mary's Harbour, July

Ranunculaceae | Buttercup Family

The oval berries look like blue beads at the end of their stems. They are poisonous to humans and could cause serious illness if eaten. But birds feed on these many-seeded berries. In fact, they are responsible for spreading the seeds to new spots where more of these beautiful lilies can grow.

This wild lily of North American woodlands has a wide range. It extends from central Labrador and the island of Newfoundland in the east to Manitoba in the west. It grows south to Michigan and the mountainous areas of North Carolina and Tennessee in the United States. But it grows most favourably in northern areas. Its scientific name *borealis*, "of the north," tells us so.

"Remember to look for the crown of five small triangular points (sepals) at the end of blueberry fruits to avoid confusing them with the shiny blue poisonous fruits of bluebead lily which have only a shallow dimple at the end."

— *Sue Meades (botanist), 2024*

Red baneberry

Forteau, July

Forteau

"I once came upon a red baneberry with white berries; it was growing in the spray zone of Churchill Falls. Botanists call a plant that has a colour different than its usual colour a 'form.' The white form of the red baneberry is called *neglecta*, which means 'neglected.'"

— *Author's Notes, 1984*

RED BANEBERRY

The berry whose name means "trouble"

 *Actaea rubra*

Glossy red berries, each one with a tiny black eye and pin-like stem, stick straight out from their clusters. They seem to be staring at us, luring us to look closer into their shady niche. But the plant's name, red baneberry, gives us fair warning. "Bane" is an old word that means trouble, harm, poison, or death.

Indeed, this plant is deadly poisonous. All its parts contain a toxic oil which is common to the buttercup family. Its fruit and roots have the highest concentrations of the oil. Eating just a few red baneberries may cause heart failure or other symptoms that are fatal.

Several species of birds and small mammals go unharmed from feeding on the red baneberry. Robins, partridges, and other birds eat the fruit, while small mammals such as squirrels and mice eat the seeds and spit out the flesh.

We find the red baneberry in moist places near lakes and streams, in mixed forests and shady thickets. It grows up to one metre in height with compound leaves of many sharp-toothed leaflets. In early summer, fragrant, white flowers open in a beautiful cluster at the top of the stem. The flowers have so many stamens that the cluster resembles lace. When the fruit develops, the blossom's stigma persists as a black dot or "eye" at the tip of the berry. Inside the berry are nine to sixteen red-brown, wedge-shaped seeds.

The red baneberry is found in the Straits and Lake Melville areas, and Labrador West. When we see its lovely, lace-like flower cluster or its red berry with alluring shine and "black-dot eye," we will know what it is. And we will do well to remember what's in its name.

Dogberry

"Dry in the house. Eat when you have a cold. Put berries in a bottle. Brew for cough syrup."
— *Kitora Boase, Hopedale, 1984*

"Best after the frost. Make jelly or roast them plain."
— *Aunt Flo Baikie, North West River, 1972*

"Boil wood and bark for a drink."
— *John Michelin, North West River, 1972*

DOGBERRY

A feast in its crown

 dogwood showy mountain-ash mashkumin *Sorbus decora*

The dogberry is well-known throughout Labrador — near the Coast in the reach of salt-spray, by streams and lakesides, on rocky slopes, and in protected places of the tundra. It grows as a shrub of many stems or small tree up to a height of seven metres. Its greyish-brown bark is marked with lenticels (lines of slightly raised pores).

> "Called dogwood. The steeped berries are medicine for TB and sore throats. Chew the berry for throat."
> — Muriel Andersen, Makkovik, 1984

The dogberry's compound leaves have eleven to fifteen finely toothed leaflets. Its white flowers are crowded into flat-topped clusters. They bloom after the leaves have appeared in spring.

Its shiny orange-red berries ripen in autumn and persist on the tree all winter.

Known for its beauty, the dogberry or mountain-ash has been planted as an ornamental in lawns, parks, gardens, and along streets for centuries. And in ancient times, the rowan tree, a mountain-ash of the British Isles and northern Europe, was planted for its symbolism and mythical powers. The worldwide distribution of mountain-ash species is impressive — over one hundred species in the north temperate areas of North America, Europe, and Asia. Here in Labrador, we see the dogberry, transplanted from the wild, brightening the yard of many a dwelling. Its species name is fitting — "*decora*," Latin for "handsome" or "beauty."

> "Boil the berry and the skin of the stick for a medicinal drink. It is the best medicine for a headache, sore throat, or chest infection."
> — An Elder in Davis Inlet, 1984

Though the tree has the striking beauty of flower cluster, berry, and autumn leaf, the berry has a disappointing taste. Bitter and unpleasant, it

Dogberry

seems fit only for dogs, hence the name "dogberry." But this has been proven otherwise in Labrador.

Widely used for medicine, its rind (bark) and berries have been boiled and steeped to cure coughs, colds, flu, and chest infections. Its berries have been eaten raw for sore throats or for restoring the appetite. And it has been gathered after the frost for jelly and wine, when the taste of the berry is greatly improved.

> "Make wine. Put in water with yeast and sugar. Let it brew."
> — Bessie Flynn, Forteau, 1984

The dogberry is also important to many species of birds. When the fruit ripens orange-red, Pine Grosbeaks or Cedar Waxwings will fill a tree and eat until the berries are gone. Canada Jays, Eastern Robins, and White-winged Crossbills feed on them as well. And animals aren't left out. Moose, along with arctic and snowshoe hare, browse on twigs and foliage; beaver feed on the bark.

As the dogberry's range extends north along the Labrador Coast, it is identified as *Sorbus decora*, var. *groenlandica** (of Greenland), a smaller variety of the species. Named thus, for it is found growing deep in the fjords of the southern tip of Greenland. Its species name "*decora*" — beauty — follows it wherever it goes.

The dogberry —
white in summer
when grey-legged stems
lift up flowers
in hundreds of clusters
of crown.

The dogberry —
orange-red in autumn
when grey-legged stems
bend down berries:
too many to hold
for a crown.

The dogberry —
many-coloured in winter
when grey-legged stems
shimmer with birds
that come for a feast
in its crown.

---

* "Pocket sites" of this variety are also found in Newfoundland.

PEAR
Petal snow

L plum **NA** chuckley pear, juneberry, saskatoon berry, serviceberry, shadbush **IA** atumin **S** *Amelanchier bartramiana*

Forteau, July

In June, when Labrador's winter snow is gone, another snowfall comes. Petal snow. We see this white-blossom snow on tall pear shrubs that are scattered along the edges of woods, peat bogs, and roadsides. We see it on smaller pear plants in protected places near the Coast. Soon the delicate flowers let their petals fall: soft, white flakes, five petals from each flower, detaching with the slightest wind.

> "The pear is one of the prettiest plants. It has showy white flowers in late spring and early summer. The fruit is bigger than most berries and a beautiful blue. In the fall the foliage turns into an array of pinks, blues and purples. It grows along the access road and is easily transplanted."
> — Eva Luther, St. Lewis, 2020

Pear

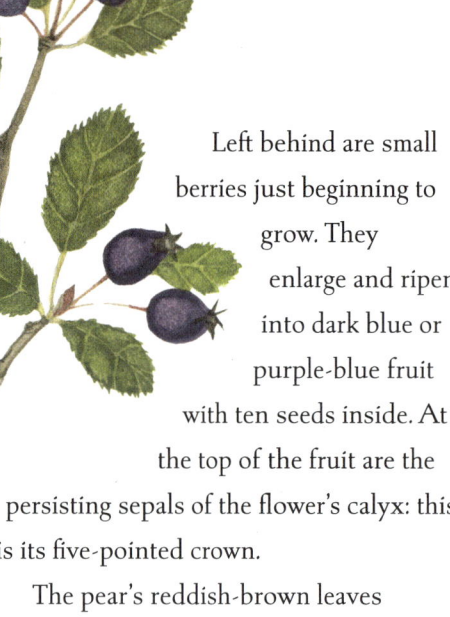

Left behind are small berries just beginning to grow. They enlarge and ripen into dark blue or purple-blue fruit with ten seeds inside. At the top of the fruit are the persisting sepals of the flower's calyx: this is its five-pointed crown.

The pear's reddish-brown leaves emerge from their own buds at the same time as the flowers emerge from theirs; they will soon turn to summer green. The colour of the leaves, the shape of the flower petals, and the size and shape of the shrub vary greatly, depending on where the plant grows.

The ripe berries are sweet and good to eat if we can pick them before the songbirds, Canada Jays, ravens, red squirrels, foxes, and black bears. This is a berry of many pickers. It is also a berry of many names.

On the prairies it's saskatoon berry;
in Newfoundland, chuckley pear.
In some places, serviceberry or shadbush,
but in Labrador it's plum or pear.

Note:
Many native species of the *Amelanchier* genus grow in North America. In fact, there is not a province nor territory nor state that is without a species of serviceberry. No one knows the exact number of species. Some sources say sixteen. Some say twenty-four. Some say more than thirty. Because serviceberries commonly hybridize, it makes them difficult to identify. But in Labrador, there is no problem in identifying our species. There is only one, *Amelanchier bartramiana*. And that is our pear!

WILD CHERRY

We call them "pioneers"

NA pin cherry, fire cherry   **IA** upueiminan   **S** *Prunus pensylvanica*

Wild cherries appear where a fire has burned the forest. They also grow by airstrips and along roads where bulldozers have kicked up the soil.

> "The cherry is a single standard, and I believe very scarce; for I met with them by the side of one hill only where they stood in good plenty… the fruit was small, tasteless, and nearly all stone."
>
> — *Captain Cartwright's journal, October 4, 1786*

Wild cherries like disturbed, open ground. Their hard, round seeds wait in the soil for the right conditions, sometimes as long as a hundred years! They don't like shade, so when a forest is cut down or burned, they germinate and grow up quickly. In only two years, the young trees will produce fruit. We call them "pioneers" — trees that make way for other trees to grow.

Rosaceae | Rose Family

Wild cherry

Happy Valley-Goose Bay, June

Their roots are also important in preventing soil erosion. They spread, holding tightly to the soil, producing root sprouts that grow up into new trees.

Wild cherries may reach a height of eight metres, but have a lifespan of not much more than thirty years. The bark is smooth and reddish-brown, marked with orange-coloured pore lines called lenticels. The white, five-petaled flowers appear in small clusters with the new leaves of spring. The sharp-pointed leaves turn lovely shades of orange and dark red in the fall.

> "I observed many wild cherries scattered along the sandy banks of the lower Grand River."
> — *Author's Notes, 1992*

> "We pick them for jelly."
> — *Sam Learning, Paradise River, 1984*

The bright red fruit hang down from long, pin-like stems — hence the name, "pin cherry."

Wild cherry trees produce an abundance of fruit. Plenty for flocks of songbirds and plenty for batches of bright red jelly. And plenty on the ground for foxes and black bears and many species of small mammals. Wild cherries are also planted as ornamentals in yards and along the road. Everyone enjoys a cherry tree!

Rosaceae | Rose Family

WILD STRAWBERRY

Freckles on its berries

 common strawberry  appiujak  uteiminan  *Fragaria virginiana*

Sweet, sweet strawberries, ripened soft, juicy, and fragrant. Staining our fingers red when we pick them. Who can find strawberries in Labrador? Captain Cartwright found them over two hundred years ago and wrote in his journal, "I saw tolerable plenty of scarlet strawberries in L'ance a Loup [*sic*] which were the only ones observed in that country."

We can find them, too, in unexpected places — near a lighthouse, by an old store, on an island, under a clothesline, in tall grasses, or in a sandy backyard. We see the little strawberry plants with their leaves of three sharp-toothed leaflets. We see their runners, thin stolons that travel along the ground to take root and make daughter plants.

Strawberry plants bear small clusters of white, five-petaled flowers. Because the flowers depend upon bees, flies, ants, and butterflies for pollination, not all flowers will become berries.

The strawberry fruit develops in an unusual way. The place where the flower grows from its stem enlarges into a berry — red outside, white inside. We can see its seeds; they are called achenes and appear as freckles on the berry's pulp. The hull of the strawberry is the flower's calyx of five green sepals.

Sweet, sweet strawberries have an old name and a long history. Most sources say that their

Rosaceae | Rose Family

Wild strawberry

common name comes from the Old English word "strew," because the runners spread and appear "strewn" like straw upon the ground. The wild strawberry of Labrador, *Fragaria virginiana*, also has a long history. It is this North American species that was first cultivated by Indigenous peoples in eastern United States.

In the 1600s, when early settlers arrived in Massachusetts, one settler described it this way: "In some parts where the Indians have planted, I have many times seen as many as could fill a good ship, within few miles compass" (Roger Williams, 1643).* Through the centuries, *Fragaria virginiana* has been hybridized with many other species of strawberry. Some of these hybrid strawberries are what we find in grocery stores today.

In Labrador, the wild strawberry is limited to the Straits, Southern Labrador, and Lake Melville areas. Plants that we find growing in yards have often been transplanted from strawberry patches outside Labrador.

Forteau, July

---

* From *Sturtevant's Edible Plants of the World*, 282.

Rosaceae | Rose Family

BAKEAPPLE

Queen of the berries

 cloudberry    akpik, appik    shikuteu    *Rubus chamaemorus*

In his Labrador journal, Captain Cartwright called the bakeapple a "baked-apple," because its taste was "like that of a roasted apple." But long before explorers or settlers arrived on the Coast of Labrador, the Inuit had named it appik and the Innu were calling it shikuteu — names all its own.

This is the orange-yellow berry that speckles the peat bogs in the amber August sun. This is the soft, juicy berry with crunchy seeds, whose texture and taste cannot be compared with any other fruit. This is akpik, shikuteu, the bakeapple. But there is yet another name from those who know it well — "queen of the berries."

And queen she is, compelling us to walk to the farthest bog to pick all day and return the next. Never mind the flies, the tired backs, the heavy buckets home. We will have our royal bakeapples all winter long — to fill our pies, top our cakes, spread on our bread, and to spoon into our bowls.

This berry queen is both tough and tender. Tough in her rhizome trail that travels underneath

> "About berry-picking as a child, I remember climbing the big hill behind our house and picking bakeapples and going home with empty containers because I would eat them as I was picking. They taste so good and juicy. Mom would make pies with them but now I make mostly cheesecake. When I was growing up, we didn't have freezers to keep the berries good over the winter, so they were put in glass jars and topped with sugar. This would keep them good all winter."
>
> — *Pauline Elson, Cartwright, 2020*

Bakeapple

the soil throughout the bog and in her reign that extends north into the arctic. Tender in her soft white petals that are easily damaged by wind and rain or summer frosts.

When there's a thunderstorm or a stretch of rainy, cold days at the time of blossoming, we hear people say, "Not many bakeapples this year!"

Male and female flowers appear on separate plants. If the petals are damaged, no pollinating insects will visit and there will be no fruit. But the bakeapple is tough again in her rhizomes, which are still able to produce new plants.

> "My family was picking bakeapples on the head of Cape Harrison, right up steep from the water. The side of the hill was spotted orange with berries. And there beside us was the ocean spotted white with icebergs. We counted them — 110 icebergs in the Labrador Current. We couldn't count the bakeapples."
>
> —*Author's Notes, 1985*

Bakeapple

"I've only come to love bakeapple picking in the last ten years or so — prior to that I couldn't understand how someone could subject themselves to the agonizing walks across barren after barren and near back-breaking work required to fill a bucket with berries. Now I love every minute of it. Walking the old footpaths, getting to spend time on the land, and the feeling of gifting bakeapples to folk who aren't fortunate to have them growing right in their 'backyard' — there's no other feeling like it really. After I pick my last bakeapple in August, I start counting the days until I can get back at it next season — picking berries for my sons and my friends who love the tart, apricot-like taste of the most spectacular berry mother nature affords us all."

— *Jamie Pye, Forteau, 2024*

Bakeapple

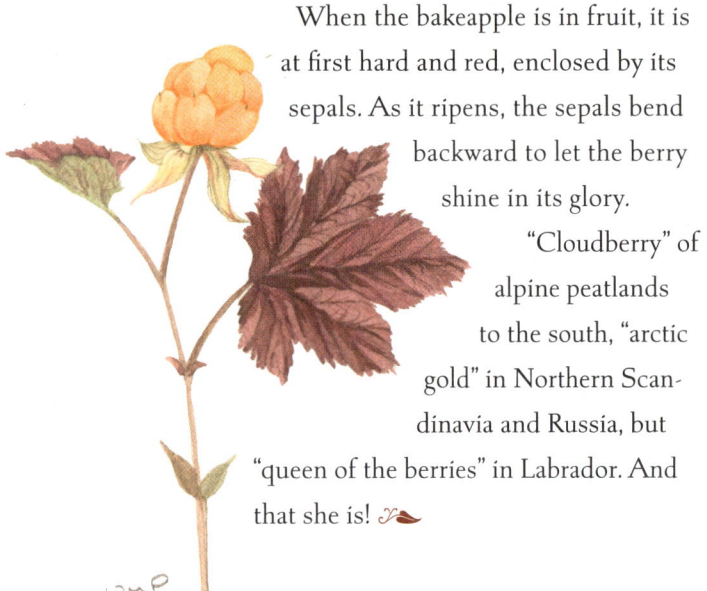

When the bakeapple is in fruit, it is at first hard and red, enclosed by its sepals. As it ripens, the sepals bend backward to let the berry shine in its glory.

"Cloudberry" of alpine peatlands to the south, "arctic gold" in Northern Scandinavia and Russia, but "queen of the berries" in Labrador. And that she is!

"When I was growing up in Black Tickle, every year on the 15th of August, which was honoured as Virgin Mary's birthday, my father would not go fishing but instead cleaned the bottom of his boat and in the afternoon took us all to the bight known as 'Trout Brook.' There we picked about five gallons of bakeapples, enough for the winter.

My parents used to make a very special meal using the bakeapples. They would cook salt beef and make doughboys which were placed in the pot for ten minutes until ready to eat. They made bakeapple jam and after all was prepared, we would eat the beef and doughboys with the bakeapple jam. Even today this is still a family favourite."

— *David Dyson, Black Tickle, 2024*

PLUMBOY
Plump little berry prince

 arctic raspberry, arctic strawberry  stemless raspberry
 akpiujak, appiujak *Rubus arcticus*, subspecies *acaulis* (stemless)

Red Bay, July

There is a small gem of a plant with hardly any stem that hides in the mosses of the peat bog, in damp thickets, or among the rocks above the coastal landwash. It is the plumboy, a sweet dwarf raspberry and a member of the rose family. Its fruit of juicy drupelets is deep red, fragrant, and as delicious as any raspberry. Its flower has five to seven petals and can be many shades of pink, as pretty and perfumed as any rose.

The plumboy is prized in many northern places and alpine areas around the world. In fact, in Russia its name means "the berry of princes." In Labrador we can say it is the "prince of berries." In the days of the cod fishery, the plumboy grew in the fish-manured bawns. It bloomed and bore berries everywhere among the rocks and under the flakes where cod were spread out to dry. It still appears in the old sites near fishing stages just above the landwash — a small gem of a berry and of Labrador history.

PLUMBOY

*Plump little
prince of berries,
rosy-cheeked,
rooted
in rich-soiled
bawns,
under flakes
of cod
drying in the sun,
growing
ripe
in favour,
in days
of royalty
when cod was king.*

Rosaceae | Rose Family

WILD RASPBERRY

We taste summer

 anushkaniminan    *Rubus idaeus*

North West River, September

The wild raspberry likes disturbed habitats — roadside clearings, logging areas, cabin sites, and the edges of forests — so that's where we go to find raspberries. Our first pick is easy; we taste summer and it is sweet. But the raspberry patch is a scratchy place. Prickly canes against our arms and legs. Old canes in the way. Sometimes it takes quite a while to get a container full of berries.

Raspberry canes are biennial, living only two years. The first-year canes are soft and bear only leaves. Second-year canes are woody and covered with bristles and sharp prickles. They bear the flowers and berries, then die. These dead canes persist, sticking up in the raspberry patch. The plant's rhizomes continue to travel underneath the soil to sprout new canes. This is the way the raspberry patch spreads. Birds flit and feed on the berries and peck at fallen fruit in its shade. Black bears come and we come, too. Sometimes we find each other there at the same time.

The flowers of the wild raspberry have five white petals and no fragrance. They hang down from small clusters in the leaf axils. The green sepals are longer than the petals and bend backward.

The fruit is composed of many little drupelets. Inside each juicy, red drupelet is a seed. When we pick a raspberry, the white end of the flower stem, the receptacle, remains behind.

Rosaceae | Rose Family

Wild raspberry

Wild raspberries are not newcomers to Labrador. In 1777, Captain Cartwright described them in his journal, "By the sides of the river (North River)… an abundance of good raspberries, both red and white, as ever I ate in my life." (The white form is known as *alba*).

Yes, the wild raspberry is common in Labrador. In fact, with more disturbed sites, more and more raspberry patches appear. How far north does it grow? Some people may know, but if they do, they may be keeping it a secret.

> "I know that there were raspberry patches on the hills north of Nain, but the exact location of these was a closely kept secret."
> — *Doris Peacock, Nain (1940–57), 1984*

> "The deeper one goes inside the bays, the more one finds."
> — *Frieda Hettasch, Hebron (1945–59), 1984*

> "I picked them (white raspberries) home. Scatter few amongst the red. We called them 'golden raspberries.'"
> — *Amy Peddle, St. Lewis and Mount Pearl, 2021*

> "One day my granddaughter, Amelia, came home with a bowl of mixed berries she had picked with a neighbor in Tub Harbour. In the bowl were white raspberries; it was the first time I ever saw them, but several people from St. Lewis have picked them here."
> — *Eva Luther, St. Lewis, 2021*

Port Hope Simpson, July

Rosaceae | Rose Family

Hairy dwarf raspberry

Red Bay, July

"We call 'em travellers."
— *Kitora Boase, Hopedale, 1984*

Rosaceae | Rose Family

HAIRY DWARF RASPBERRY

Running up and over and in between

 dewberry   anikutshauminan   *Rubus pubescens*

North West River, September

When we walk in disturbed open forest places, brookside thickets, or along gravelly shores, we often come upon the hairy dwarf raspberry. Unlike the common wild raspberry with its tall, prickly, woody canes, this raspberry has smaller, slender, slightly woody stems. The plant is known for its horizontal stolons called "runners." These "runners" travel along the ground, up and over stumps, and in between rocks. Nothing stops them! Their long, tail-like tips send down roots to start new plants as they go.

Both the stems and the leaves of the hairy dwarf raspberry have tiny hairs. They are hard to see, but they are there. The leaves have three-leaflets. The flowers have five to seven white or pink petals.

The fruit is dark red or purplish-black. It is sweet and tasty, but hard to pick. It drops off late or often doesn't come off its receptacle at all. At the end of summer we often see the once-juicy fruit of the hairy dwarf raspberry all dried up, still on its stem.

Rosaceae | Rose Family

POISONBERRY

Secretly stealing food and drink

 L spiderberry   **NA** northern comandra, false toadflax   **S** *Geocaulon lividum*

Here an orange berry, there an orange berry in the redberry patch. They look poisonous, so we call them poisonberries. And yes, to eat just a few could be harmful. So, we don't even try them.

Poisonberry flowers appear in early summer, two to four tiny yellowish, greenish, purplish flowers in the leaf axils midway up the plant's stem. The flowers have no petals but five triangular sepals. Only one of these flowers develops into a berry — a juicy, orange-red berry with a large pit inside. The leaves are green, lead-coloured, or tinged with purple.

What is the poisonberry doing in the redberry patch? Beneath the ground, it hides long, creeping rhizomes. These rhizomes have thin threads that attach themselves to the roots of plants and steal some of their food and water. The poisonberry is semi-parasitic, stealing nutrients from redberry roots. It is also semi-parasitic on the roots of other Labrador plants, including the foxberry, blackberry, blueberry, and crackerberry.

Poisonberry rhizomes travel a long way just underneath the surface, producing new plants here and there along its length. That's why we can say, "Here an orange berry, there an orange berry in the redberry patch!"

The poisonberry is found in peat bogs, berry patches, amidst shrubbery, and at the edges of forests throughout Labrador north to the Nain area.

North West River, September

76   Santalaceae | Sandalwood Family

Poisonberry

"We call 'em spiderberries."
— *John Michelin, North West River, 1972*

In its wider range outside Labrador, the plant is called northern comandra or false toadflax, extending west to Alaska and south to the boreal forests of Montana. In the east it appears south into northern New England and New York State, where it is uncommon or rare.

Mary's Harbour, July

Santalaceae | Sandalwood Family

ILLUSTRATED BOTANICAL GLOSSARY

With Illustrations by Mavis Penney

Many botanical terms are used in this book; however, everyday language is preferred whenever possible. For example, all different types of fruit are called "berries," because that is how they are known in Labrador. The following glossary illustrates several botanical terms that appear in the plant descriptions, while many more terms explain themselves in the text.

Illustrated Botanical Glossary

Detailed definitions of botanical terms		Illustrated diagrams
SIMPLE LEAF	One leaf blade whose petiole (stalk) attaches the leaf to the stem.	petiole
COMPOUND LEAF	A leaf made up of two or more leaflets.	petiole, leaflet
BASAL ROSETTE	Basal leaves: leaves growing from the base of the stem. Basal rosette: basal leaves making a circle around the plant.	basal leaf, basal rosette
AXIL	The space (angle) between leaf and stem, the place where they meet.	axil

Berries of Labrador

Illustrated Botanical Glossary

STOLON A stem that runs along on top of the ground from the base of a plant, making "daughter plants" as it goes. A stolon is often called a "runner."

RHIZOME A stem that extends horizontally underneath the ground, sprouting up new "daughter plants" as it goes.

RECEPTACLE Receptacle: the place at the base of the flower where the fruit or seed is attached.

Calyx: the sepals of the flower. The calyx often remains when the flower has become fruit. In the strawberry and raspberry, the persisting calyx is known as the "hull."

SPATHE

SPADIX

Spathe: the hood in the flowers of the Arum Family. This hood is actually a bract, a leaf that belongs to the flower.

Spadix: a spike of many flowers.

BRACT Petal-like leaves that belong to the flower, or tiny leaves that appear on other parts of the plant.

Illustrated Botanical Glossary

PERFECT FLOWER

Perfect flower: a flower that has both male and female parts in one blossom.

Corolla: the petals together are called the corolla.

Calyx: the leaf-like sepals under the corolla together are called the calyx.

Sepals: the leaf-like parts that make up the calyx under the flower.

Pistil: female part of the flower, composed of stigma, style, and ovary (seeds develop in the ovary).

Stamen: male part of the flower composed of anther and filament (pollen is stored in the anther).

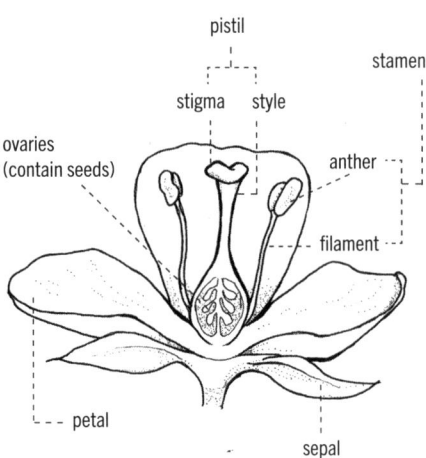

Other Botanical Terms Not Included In This Diagram:

Imperfect flower: a flower that has only male or only female parts in its blossom.

Tepals: petals and sepals of similar shape and colour that are hard to tell apart. The species in this book that have tepals are: twisted-stalk, Canada mayflower, yellow clintonia, and three-leaved false Solomon's seal.

INTRODUCTION TO FRUIT TABLE

As we look closely at the berries, comparing them with each other, we see differences in how the berries protect or "make houses" for their seeds. Botanists call all of what we call berries "fruit." Then they place each fruit in a special group based on how they protect or "house" their seed. The fruit table tells the name of the group to which each fruit belongs (not all fruits fall into distinct categories). There are seven kinds of fruit in the fruit table.

1. Achene — small, hard seeds in little pits on the outside of the fruit (e.g. strawberry)
2. Berry — fleshy or juicy fruit containing many small seeds (e.g. blueberry)
3. Berry-like capsule — many tiny seeds in a dry capsule, enclosed by the sepals of the flower which become fleshy like a berry (e.g. maidenhair berry)
4. Drupe — fleshy fruit with a seed enclosed in a stone or nutlet (e.g. foxberry)
5. Drupelets — little drupes in a cluster (e.g. raspberry)
6. Pome — fleshy fruit with a tough core around the seeds (e.g. dogberry)
7. Cone — seed fruit of a conifer (e.g. common juniper)

> **IMPORTANT NOTE**
>
> **Edible/Inedible:** These two categories represent "what we eat" and "what we don't eat" in Labrador. This is based on a long tradition of berry-use by the people who live close to the land.

Berries of Labrador

LABRADOR FRUIT TABLE

Page #	Scientific Name	Common Name(s)	Fruit	Colour	Edibile or Inedible
56–57	Actaea rubra	red baneberry	berry	cherry-red	inedible, poisonous
	Actaea rubra form neglecta			white	inedible, poisonous
61–62	Amelanchier bartramiana	pear, serviceberry	pome	reddish purple or dark blue	edible, sweet
10–11	Aralia hispida	bristly sarsaparilla	drupe	dark purple to black	inedible, considered poisonous*
30–31	Arctostaphylos uva-ursi	bearberry	drupe	red	edible, dry, mealy
28–29	Arctous alpina	foxberry, alpine bearberry	drupe	shiny red to purplish black	edible, juicy, lacks flavour
8–9	Calla palustris	wild calla	berry	scarlet red	inedible, poisonous
54–55	Clintonia borealis	bluebead lily, yellow clintonia	berry	shiny blue	inedible, poisonous
17–19	Cornus canadensis	crackerberry, bunchberry	drupe	bright red	edible, pulpy, lacks flavour
20–21	Cornus sericea	red-osier dogwood	drupe	white, often with bluish tint	edible, very bitter
18	Cornus suecica	Swedish crackerberry	drupe	bright red	edible, sweeter and juicier than C. canadensis †
27	Empetrum atropurpureum	purple crowberry	drupe	dark red to reddish purple	edible, lacks flavour

† From Merritt Lyndon Fernald, *Gray's Manual of Botany*, 8th edition.

Page #	Scientific Name	Common Name(s)	Fruit	Colour	Edibile or Inedible
27	*Empetrum eamesii*	pink crowberry, rockberry	drupe	pink to light red	edible, lacks flavour
24–27	*Empetrum nigrum*	blackberry, crowberry	drupe	raven-black	edible, watery, sweet
65–66	*Fragaria virginiana*	wild strawberry	achenes pitted in fleshy receptacle	red	edible, very sweet
32–33	*Gaultheria hispidula*	maidenhair berry, creeping snowberry	berry-like capsule	white with tiny, soft bristles	edible, wintergreen flavour
76–77	*Geocaulon lividum*	poisonberry, northern comandra	berry-like drupe	bright orange-red	inedible, considered poisonous*
22–23	*Juniperus communis*	ground juniper	berry-like cone (1–3 seeds inside)	blue with chalky bloom	edible, aromatic
16	*Lonicera villosa*	northern honeysuckle	berry	bluish-black	edible, juicy, sweet
12–13	*Maianthemum canadense*	Canada mayflower	berry	speckled when unripe, turning to crimson	inedible, considered poisonous*
15	*Maianthemum stellatum*	starry false Solomon's seal	berry	green, striped with black, then dark red	inedible, considered poisonous*
14–15	*Maianthemum trifolium*	three-leaved false Solomon's seal	berry	green with reddish spots, turning red	inedible, considered poisonous*

Berries of Labrador

Labrador Fruit Table

Page #	Scientific Name	Common Name(s)	Fruit	Colour	Edibile or Inedible
63–64	Prunus pensylvanica	wild cherry	drupe	bright red	edible, sour; stone poisonous
47–48	Ribes glandulosum	currant, red currant	berry	red with soft, bristly hairs	edible, tart
50–51	Ribes lacustre	bristly black currant	berry	purplish-black with gland-tipped bristles	edible, very sweet
49	Ribes triste	swampy red currant	berry	red, lacking bristles	edible
71	Rubus arcticus subsp. acaulis	plumboy, arctic raspberry	cluster of drupelets	deep red	edible, fragrant, very sweet
67–70	Rubus chamaemorus	bakeapple, cloudberry	cluster of drupelets	orange, amber	edible, juicy, a taste of its own
72–73	Rubus idaeus	wild raspberry	cluster of drupelets	red	edible, sweet
74–75	Rubus pubescens	hairy dwarf raspberry, dewberry	cluster of drupelets	dark red to purplish-black	edible
58–60	Sorbus decora	dogberry	pome	orange-red	edible, bitter
52–53	Streptopus amplexifolius	twisted-stalk, clasping-leaved twisted-stalk	berry	red	inedible, considered poisonous*.
53	Streptopus lanceolatus	rose twisted-stalk	berry	red	inedible, considered poisonous*

Labrador Fruit Table

Page #	Scientific Name	Common Name(s)	Fruit	Colour	Edibile or Inedible
40–41	*Vaccinium angustifolium*	blueberry, lowbush blueberry, tobacco hurts	berry	dark blue with bloom	edible, sweet
42	*Vaccinium boreale*	northern blueberry, tobacco hurts	berry	dark blue with bloom	edible, varies in sweetness
45–46	*Vaccinium cespitosum*	sugar hurts, dwarf bilberry	berry	light blue	edible, very sweet
34–35	*Vaccinium oxycoccos*	marshberry, small cranberry	berry	speckled then dark red	edible, tart, sweeter in spring
43–44	*Vaccinium uliginosum*	ground hurts, blueberry, alpine bilberry	berry	deep blue, whitish bloom	edible
36–38	*Vaccinium vitis-idaea*	redberry, partridgeberry, mountain cranberry	berry	dark, shiny red	edible, tart, sweeter in spring
6–7	*Viburnum edule*	squashberry	drupe	orange-red	edible, tart

* Indicates that other book or online sources may give the fruit a different edibility rating.
See: *Plants for a Future*, https://pfaf.org/

Berries of Labrador

ACKNOWLEDGEMENTS

The author would like to thank the following:

Dr. Peter J. Scott, who edited the original manuscript in 1985.

Canada Council Explorations Grant, which funded research on the Labrador flora in 1984.

Labrador Institute, publisher of the 2021 edition of *Berries of Labrador*.

Dr. Ashlee Cunsolo, **Morgen Mills**, **Dr. Alexandra Sawatzky**, and **Dr. Erica Oberndorfer**, for editing, design, production, and review of the 2021 edition.

Eva Luther, for her cultural review of the manuscript and for providing numerous contacts and information on local uses and names of the berries.

Isabel Watts and **Liz Dawson**, for their help with research on the distribution and local uses of the berries in the Lake Melville area.

Gerry Bance, for his phytogeographic contributions and edits.

Susan Meades, for her botanical field information and edits.

Sarah Townley, for her Inuttitut translation of berry names.

Marguerite MacKenzie and José Mailhot, for Innu-aimun names and uses of the berries, through their *Innu-English Dictionary*.

Mavis Penney, for her illustrations of glossary terms and the fruit of three berry species.

Valerie Powell, for her beautiful watercolours throughout the book and for her travel to numerous places in Labrador to paint them on site.

Leah Gomes, for her technical support for the project.

Susan Rogers, my cousin, who provided many helpful Alaskan and Russian botanical sources.

Carol Bryan, my sister, for her invaluable support throughout my writing career, especially during this berry project from its beginnings in 1984 through its publication in 2024.

Memorial University Press, publisher of the 2024 edition of *Berries of Labrador.*

Alison Carr, for her work as managing editor and exquisite graphic designing for this edition; **Randy Drover**, for his marketing and sales expertise; **Vicki Hallett** and **Fiona Polack**, fine academic editors, for their eager help and support; and **Rebecca Roberts**, for her excellent work as copyeditor and the pleasure it was to work with her.

Community residents throughout Labrador, who provided information about the berries reflected in the quotes.

The late Clarice Hopkins of Cartwright and Indian Tickle, who, through her stories of berries and berry-picking, made the berries come alive to me.

MIXED BERRY JAM

"It is late September and I am on the side of Nain Hill picking berries. Colours are mixed-up on the land. Flies and snow are mixed-up about my face. My fingers sting with the cold but I am still picking."

— *Author's Notes, 1981*

BAKEAPPLE AND BLACKBERRY; REDBERRY, BLACKBERRY, AND BLUEBERRY; REDBERRY AND BAKEAPPLE; SQUASHBERRY AND REDBERRY; RASPBERRY, STRAWBERRY, AND RHUBARB; RASPBERRY AND CURRANT; BAKEAPPLE, BLACKBERRY, AND BLUEBERRY;

BIBLIOGRAPHY

Agriculture, Aquaculture and Fisheries, New Brunswick. "Wild Blueberry Fact Sheet A.2.0: Growth and Development of the Wild Blueberry." Government of New Brunswick. Revised 2010. https://www2.gnb.ca/content/dam/gnb/Departments/10/pdf/Agriculture/WildBlueberries-BleuetsSauvages/a20e.pdf?random=1720798126621

Ainana, Lyudmila and Igor Zagrebin. *Edible Plants Used by Siberian Yupik Eskimos of Southeastern Chukchi Peninsula Russia.* Anchorage, AK: National Park Service, 2014.

Appalachian Mountain Club. *Mountain Flowers of New England.* Boston: Appalachian Mountain Club, 1964.

Austin, Oliver Luther Jr. *The Birds of Newfoundland and Labrador.* Cambridge, MA: Nuttall Ornithological Club, 1932.

Baikie, Flora, Rosie Ford, Pearl Fowler, Clara Ford, and John Pardy. "Bakeapples." *Them Days* 8, no. 1 (1982): 22-25.

Bandringa, Robert W. and Inuvialuit Elders. *Inuvialuit Nautchiangit – Relationships Between People and Plants.* Inuvik, NT: Inuvialut Cultural Resource Centre, 2010.

Bishop, Harlow. "The Austin Collection from the Labrador Coast." *Rhodora* 32, no. 375 (1930): 59-62.

Blondeau, Marcel and Claude Roy. *Atlas of Plants of the Nunavik Villages.* Sainte-Foy, QC: Éditions MultiMondes, 2004.

Bibliography

Bodsworth, Fred. *Last of the Curlews*. 1955. Washington, DC: Counterpoint Press, 1995.

Boland, Todd. *Trees and Shrubs of Newfoundland and Labrador*. Portugal Cove-St. Philip's, NL: Boulder Publications, 2011.

Boland, Todd. *Wildflowers and Ferns of Newfoundland and Labrador*. Portugal Cove-St. Philip's, NL: Boulder Publications, 2017.

Brassard, G.R., Stanley Frost, Marshall Laird, O.A. Olsen, and D.H. Steele. "Studies of the Spray Zone of Churchill Falls, Labrador." *Biological Conservation* 4, no. 1 (1971): 13-18.

Burt, Page. *Barrenland Beauties: Showy Plants of the Canadian Arctic*. Yellowknife: UpHere Publishing Ltd., 2004.

Cartwright, George. *Captain Cartwright and His Labrador Journal*. Edited by Charles Wendell Townsend, M.D. 3rd edition. St. John's: DRC Publishing, 1911.

Craighead, John J., Frank C. Craighead, Jr., and Ray J. Davis. *A Field Guide to Rocky Mountain Wildflowers*. New York: Houghton Mifflin, 1963.

Cuerrier, Alan (IRBV) and Inuit Elders, Nunavik. *The Botanical Knowledge of the Inuit of Kangiqsujuaq, Nunavik*. 4th edition. Inukjuak, QC: Avataq Cultural Institute, 2005.

Cuerrier, Alan and Luise Hermanutz. *Our Plants, Our Land: Plants of Nain and Torngat Mountains Basecamp and Research Station (Nunatsiavut)*. Montreal and St. John's: Institute de recherche en biologie vegetale Jardin botanique de Montreal and Department of Biology, Memorial University, 2012.

Earl, John, Bessie Flynn, and Leslie Pardy. "Berry Picking." *Them Days* 6, no. 1 (1980): 28-31.

Bibliography

Farrar, John Laird. *Trees in Canada.* Markham, ON: Fitzhenry & Whiteside and The Canadian Forest Service, 1995.

Fernald, Merritt Lyndon. *Gray's Manual of Botany.* 8th edition. New York: American Book Company, 1950.

Forsyth, Adrian. *Mammals of North America (Temperate and Arctic Regions).* Buffalo, NY and Kingston, ON: Firefly Books and Bookmakers Press, 1999.

Foster, Steven and James A. Duke. *A Field Guide to Medicinal Plants and Herbs.* Boston: Houghton Mifflin, 1990.

Grenfell, Wilfred T., et al. *Labrador, the Country and the People.* New York: The MacMillan Company, 1910.

Harper, Francis. *Plant and Animal Associations in the Interior of the Ungava Peninsula.* Lawrence, KS: University of Kansas, 1964.

Hedrick, U.P., ed. *Sturtevant's Edible Plants of the World.* 1919. New York and Toronto: Dover Publications and General Publishing Company, 1972.

Holden, Alexander Edward. *Plant Life in the Scottish Highlands.* Edinburgh: Oliver and Boyd, 1952.

Hulten, Eric. *Flora of Alaska and Neighboring Territories: A Manual of the Vascular Plants.* Stanford: Stanford University Press, 1968.

Hustich, Ilmari. "The Introduced Flora Element in Central Quebec-Labrador Peninsula." *Canadian Field Naturalist* 98 (1971): 425-441.

Inuit Tapiriit Kanatami. "Nain – Rutie Dicker." YouTube video. November 20, 2018. https://www.youtube.com/watch?v=goqdDqMTIJY

Joamie, Aalasi and Anna Ziegler. *Walking with Aalasi: An Introduction to Edible and Medicinal Arctic Plants.* Toronto and Iqaluit: Inhabit Media, 2009.

Johnson, Karen L. *Wildflowers of Churchill and the Hudson Bay Region.* Winnipeg: Museum of Man and Nature, 1987.

Bibliography

Learning, Jemima. "Berries 'n Things." *Them Days* 1, no. 2 (1975): 20-23.

Lysaght, A.M. *Joseph Banks in Newfoundland and Labrador, 1766: His diary, manuscripts, and collections.* Berkley, CA: University of California Press, 1971.

Macoun, James Melville. *List of the Plants Known to Occur on the Coast and in the Interior of the Labrador Peninsula.* London: FB&C Ltd. (Forgotten Books), 2018.

Makinen, Yrjo. "The Most Important Plants of the Schefferville Area, Central Labrador." *Occasional Papers of the Kevo Subarctic Club, University of Turku* 1 (1978): 1-16.

Mallory, Carolyn. *Common Insects of Nunavut.* Iqaluit: Inhabit Media, 2012.

Mallory, Carolyn and Susan Aiken. *Common Plants of Nunavut.* Iqaluit: Department of Education and Canadian Museum of Nature, 2004.

Martin, Alexander C., Herbert S. Zim, and Arnold L. Nelson. *American Wildlife and Plants: A Guide to Wildlife Food Habits.* Toronto: General Publishing Company, 1951.

Meades, S.J. and L. Brouillet. "Annotated Checklist of the Vascular Plants of Newfoundland and Labrador." In *Flora of Newfoundland and Labrador.* S.J. Meades and W.J. Meades. Sault Ste. Marie, ON: 2019+. https://www.newfoundland-labradorflora.com/checklist/

Mittelhauser, Glen H., Jensen Bissell, Dan Cameron, Alison C. Dibble, Arthur Haines, Jean Hoekwater, Marilee Lovit, and Aaron Megquier. *The Plants of Baxter State Park.* Orono, ME: University of Maine Press, 2016.

Mittelhauser, Glen H., Linda L. Gregory, Sally C. Rooney, and Jill E. Weber. *The Plants of Acadia National Park.* Orono, ME: University of Maine Press, 2010.

Moreman, Daniel E. *Native American Ethnobotany.* Portland, OR: Timber Press, 1999.

Bibliography

Oliver, Ann. "A Couple a' short Ones: Me Blackberry Skirt." *Them Days* 4, no. 2 (1978): 44.

Peacock, Doris Marian. "La Flora Del Labrador Settentrionale [The Flora of a Labrador Settlement]." *IL Polo*, September 1983: 76-80.

Peacock, F.W. with L. Jackson. *Reflections from a Snowhouse.* St. John's: Jesperson Press, 1986.

Pell, Susan K. and Bobbi Angell. *A Botanist's Vocabulary.* Portland, OR: Timber Press, 2016.

Pielou, E.C. *A Naturalist's Guide to the Arctic.* Chicago and London: The University of Chicago Press, 1994.

Porsild, A.E. *Illustrated Flora of the Canadian Arctic Archipelago.* 2nd edition. Bulletin No. 146. Ottawa: National Museum of Canada, 1964.

Pratt, Verna E. *Alaska's Wild Berries and Berry-like Fruit.* Anchorage, AK: Alaskakrafts, 1995.

Rouleau, Ernest. *Rouleau's List of Newfoundland Plants.* St. John's: Oxen Pond Botanic Park, 1978.

Ryan, A. Glen. *Native Trees and Shrubs of Newfoundland and Labrador.* 1978. St. John's: Parks Division, Department of Tourism, Government of Newfoundland and Labrador, 1995.

Scott, Peter J. *Edible Plants of Newfoundland and Labrador.* Portugal Cove-St. Philip's, NL: Boulder Publications, 2010.

Scott, Peter J. *Wildflowers of Newfoundland and Labrador.* Portugal Cove-St Philip's, NL: Boulder Publications, 2006.

Shacklette, Hansford T. "Field Observations of Variation in Vaccinium uliginosum." *The Canadian Field Naturalist* 76, no. 3 (1962): 162-166.

Bibliography

Story, G.M., W.J. Kirwin and J.D.A. Widdowson. *Dictionary of Newfoundland English.* Toronto: University of Toronto Press, 1982.

Vander Kloet, S.P. "The Taxonomy of Vaccinium and Oxycoccus." *Rhodora* 85, no. 841 (1983): 1-43.

Walker, Marilyn. *Harvesting the Northern Wild.* Yellowknife: Outcrop Ltd., 1984.

Warkentin, Ian and Sandy Newton. *Birds of Newfoundland.* Portugal Cove-St. Philip's, NL: Boulder Publications, 2009.

Wells, Diana. *100 Flowers and How They Got Their Names.* Chapel Hill, NC: Algonquin Books, 1997.

NOTES ON CONTRIBUTORS

ELLEN BRYAN OBED

Raised on a small farm in Maine, Ellen Bryan Obed first went to Labrador at the age of twenty to work at Cartwright Summer Camp. After University, she returned to Labrador to teach school and to research its flora. She married Enoch Obed of Nain and they had three children, Keturah, Natan, and Seth.

Ellen is the author of ten books for children. Her *Borrowed Black* has been translated into seven languages and appeared in thirteen different editions since it was first published in 1979. The seasons and natural history provide the subjects for most of Ellen's writing. She and her husband, Robert, reside in the town of Ellsworth on the Coast of Maine.

VALERIE POWELL

Valerie Powell is a nurse and midwife from Cornwall, England who worked for the Grenfell Regional Health Services on the Labrador Coast from 1978–1989. She became interested in the wild flowers and berries whilst she was there and began painting them.

MAVIS PENNEY

Mavis Penney is a Canadian visual artist. She maintains a landscape painting blog and conducts collaborative online arts projects with other artists from around the world. She is currently working on a series of landscape paintings based on her travels in Northern Labrador.

INDEX
The Berries of Labrador

A

Actaea rubra, 57, 84
akpik, 67
akpiujak, 71
Amelanchier bartramiana, 61, 62
anikamin, 54
anikutshashimin, 12
anikutshauminan, 75
anushkaniminan, 72
appik, 67
appiujak, 65, 71
Aralia hispida, 11, 84
Arctostaphylos uva-ursi, 30, 84
Arctous alpina, 28, 84
assimin, 24
atumin, 61

B

bakeapple, 2, 67–70, 86, 91
baneberry, red, 56–57, 84
bearberry, 29, 30–31, 84
    alpine, 28, 31, 84
bilberry, 39
    alpine, 43, 44, 46, 87
    bog, 43, 44
blackberry, 24–27, 29, 76, 85, 91
bluebead lily, 54–55, 84
blueberry, 11, 16, 39, 40–41, 42, 43, 45, 55, 76, 83, 87, 91
    lowbush, 40, 41, 87
    northern, 42, 87
bristly sarsaparilla, 10–11, 84
bunchberry, 17, 84
    Swedish, 18
    Lepage's, 19

C

calla, wild, 8–9, 84
Calla palustris, 8, 84
Canada mayflower, 12–13, 82, 85
capillaire, 32
cherry
    fire, 63
    pin, 63, 64
    wild, 2, 63–64, 86
chuckley pear, 61, 62
clintonia, 54, 82, 84
Clintonia borealis, 54, 84
cloudberry, 67, 70, 86
cobbler, 28
Convallaria majalis, 12
Cornus
    canadensis, 17, 84
    sericea, 20, 84
    stolonifera, 20
    suecica, 18, 84
    ×*lepagei*, 19
crackerberry, 17–19, 76, 84
    Swedish, 18, 19, 84
    Lepage's, 19
cracker-jack, 17, 18
crackers, 17, 18
cranberry
    bog, 34
    large, 35
    lowbush, 38
    mountain, 29, 36, 38, 87
    small, 34, 35, 87
    wren's egg, 34
creeping snowberry, 32, 33, 85
crowberry, 26, 85
    black, 24
    pink, 27, 85
    purple, 27, 84
curlewberry, 24, 25
currant, 47–48, 49, 50, 86, 91
    bristly-black, 50–51, 86
    introduced black, 51

Index

skunk, 47
swampy red, 49, 86

D
dogberry, 58–60, 83, 86
dogwood, 59
    creeping, 17
    red-osier, 20–21, 84
    red-twig, 20
dwarf cornel, 18

E
Empetrum
    atropurpureum, 27, 84
    eamesii, 27, 85
    nigrum, 24, 25, 26, 27, 44, 85

F
foxberry, 29–29, 31, 76, 83, 84
Fragaria virginiana, 65, 66, 85

G
Gaultheria hispidula, 32, 85
Geocaulon lividum, 76, 85

H
hairy-berry, 47
honeysuckle
    mountain fly, 16
    northern, 16, 85
huggleberries, 39, 45, 46
hurts
    ground, 29, 39, 43–44, 46, 87
    sugar, 45–46, 87
    sweet, 39, 45, 46
    tobacco, 39, 40, 42, 87

I
inniminan, 40

J
juneberry, 61
juniper
    common, 22, 83
    ground, 22–23, 85
    low, 22
Juniperus communis, 22, 85

K
kakatshimin, 22
kakillânik, 17, 18
kakumin, 50
kallak, 28
kigutanginnak, 40, 42
kimminak, 36
kimminaujak, 47
kinnikinnick, 30
Kisittutaujak, 22

L
lily-of-the-valley, 12
    wild, 12
lingonberry, 36, 38
Lonicera villosa, 16, 85

M
magna-tea, 32
Maianthemum
    canadense, 12, 85
    stellatum, 15, 85
trifolium, 14, 85
maidenhair berry, 32, 33, 83, 85
marshberry, 34, 35, 87
mashkumin, 59
massekuminan, 34
minitshimin, 47
mint, 32
minty berry, 32
mountain-ash, showy, 59
mushumin, 7

N
northern comandra, 76, 77, 85

P
partridgeberry, 36, 38, 87
paungak, 24
pear, 2, 61–62, 84
pineminanish, 32
plumboy, 71, 86
poisonberry, 76–77, 85
Prunus pensylvanica, 63
pungajok, 43, 45

R

raspberry, 81, 83, 91
    arctic, 71, 86
    hairy dwarf, 74–75, 86
    wild, 72–73, 86
redberry, 29, 36–38, 46, 76, 87, 91
Ribes
    glandulosum, 47, 86
    lacustre, 50, 86
    triste, 49, 86
Rubus
    arcticus, 71, 86
    chamaemorus, 67, 86
    idaeus, 72, 86
    pubescens, 75, 86
    subsp. *acaulis*, 71, 86

S

saskatoon berry, 61, 62
serviceberry, 20, 61, 62, 84
shadbush, 61, 62
shashakuminan, 17
shikuteu, 67

singalâk, 17
Sorbus decora, 59, 60, 86
spiderberry, 76
squashberry, 6–7, 87, 91
starry false Solomon's seal, 15, 85
strawberry, 81, 83, 91
    arctic, 71
    common, 65
    wild, 2, 65–66, 91
Streptopus
    amplexifolius, 52, 86
    lanceolatus, 53, 86

T

three-leaved false Solomon's seal, 14–15, 82, 85
twisted-stalk, 52–53, 82, 86
    clasping-leaved, 52, 86
    rose, 53, 86

U

uapushimin, 11
uishatshimin, 36
upueiminan, 63

uteiminan, 65
utshishteshu, 8

V

Vaccinium, 39, 46
    angustifolium, 40, 87
    boreale, 42, 86
    cespitosum, 45, 87
    macrocarpon, 35
    oxycoccos, 34, 87
    uliginosum, 43, 87
    vitis-idaea, 36, 87
Viburnum edule, 7, 87

W

water arum, 8
water-dragon, 8
whortleberry, 46
    bog, 46
    ground, 46
    red, 38, 46
wintergreen, 32, 33

Original linocut created by Ellen Bryan Obed in 1967, transferred in 2020.